Communications
in Computer and Information Science 1521

More information about this series at https://link.springer.com/bookseries/7899

Ljupcho Antovski · Goce Armenski (Eds.)

ICT Innovations 2021

Digital Transformation

13th International Conference, ICT Innovations 2021
Virtual Event, September 27–28, 2021
Revised Selected Papers

 Springer

Editors
Ljupcho Antovski ⓘ
Ss. Cyril and Methodius University in Skopje
Skopje, North Macedonia

Goce Armenski ⓘ
Ss. Cyril and Methodius University in Skopje
Skopje, North Macedonia

ISSN 1865-0929   ISSN 1865-0937 (electronic)
Communications in Computer and Information Science
ISBN 978-3-031-04205-8   ISBN 978-3-031-04206-5 (eBook)
https://doi.org/10.1007/978-3-031-04206-5

This Springer imprint is published by the registered company Springer Nature Switzerland AG
The registered company address is: Gewerbestrasse 11, 6330 Cham, Switzerland

Preface

The ICT Innovations conference series has established itself as an international forum for presenting scientific results related to innovative fundamental and applied research in ICT. The conference aims to bring together academics as well as industrial practitioners to share their most recent research, practical solutions, and experiences and to discuss the trends, opportunities, and challenges in the field of computer science and engineering.

The 13th ICT Innovations conference (ICT Innovations 2021) took place online during September 27–28, 2021. This year, the conference theme was "Digital Transformation". Digital Transformation is the adoption of digital technology to transform services or businesses, through replacing non-digital or manual processes with digital processes or replacing older digital technology with newer digital technology. Digital solutions enable new types of innovation and creativity, rather than simply enhancing and supporting traditional methods.

In the past challenging year, on average, digital solutions have leapfrogged several years of progress in a matter of months. We are witnessing accelerated transformations in the areas of, just to name a few, e-commerce, software systems, networking, telecommunications, marketing, banking, finance, fintech, blockchain and digital leger technologies, artificial intelligence, augmented reality, transportation and delivery, and health care and telehealth. Along the way there have been many challenges, but also innovative solutions.

Conference topics of interest included, but were not limited to, the following:

- Digital transformation technologies
- All topics relating to computer science and engineering

The conference consisted of regular sessions with technical contributions (regular papers), which were single-blind reviewed and selected by an international Program Committee, as well as invited talks presented by leading scientists. Four different workshops were held in line with the main conference. This year there were 58 submissions and 15 papers were selected, giving an acceptance rate of 25.8%, with an average of 3.6 reviews per paper received.

September 2021

Ljupcho Antovski
Goce Armenski

Organization

General Chairs

Ljupcho Antovski	Ss. Cyril and Methodius University, North Macedonia
Goce Armenski	Ss. Cyril and Methodius University, North Macedonia

Scientific Committee

Ljupcho Antovski	Ss. Cyril and Methodius University, North Macedonia
Goce Armenski	Ss. Cyril and Methodius University, North Macedonia
Danco Davcev	Ss. Cyril and Methodius University, North Macedonia
Dejan Gjorgjevikj	Ss. Cyril and Methodius University, North Macedonia
Boro Jakimovski	Ss. Cyril and Methodius University, North Macedonia
Sasho Gramatikov	Ss. Cyril and Methodius University, North Macedonia

Technical Committee

Dimitar Kitanovski	Ss. Cyril and Methodius University, North Macedonia
Ana Todorovska	Ss. Cyril and Methodius University, North Macedonia
Jovana Dobreva	Ss. Cyril and Methodius University, North Macedonia

Program Committee

Aleksandar Jevremović	Singidunum University, Serbia
Aleksandra Dedinec	Ss. Cyril and Methodius University, North Macedonia
Aleksandra Mileva	Goce Delcev University of Stip, North Macedonia
Aleksandra Popovska-Mitrovikj	Ss. Cyril and Methodius University, North Macedonia

Amjad Gawanmeh	University of Dubai, UAE
Ana Madevska Bogdanova	Ss. Cyril and Methodius University, North Macedonia
Andrej Brodnik	University of Ljubljana, Slovenia
Andrej Grgurić	Ericsson Nikola Tesla d.d., Croatia
Andreja Naumoski	Ss. Cyril and Methodius University, North Macedonia
Anirban Kundu	Netaji Subhash Engineering College, India
Antonio De Nicola	ENEA, Italy
Antun Balaz	Institute of Physics Belgrade, Serbia
Arianit Kurti	Linnaeus University, Sweden
Augostino Marengo	University of Bari, Italy
Bekim Fetaji	Mother Teresa University, North Macedonia
Betim Cico	EPOKA University, Albania
Biljana Mileva Boshkoska	Jožef Stefan Institute, Slovenia
Biljana Tojtovska	Ss. Cyril and Methodius University, North Macedonia
Blagoj Ristevski	University "St. Kliment Ohridski" - Bitola, North Macedonia
Bojana Koteska	Ss. Cyril and Methodius University, North Macedonia
Boris Delibašić	University of Belgrade, Serbia
Boris Vrdoljak	University of Zagreb, Croatia
Bryan Scotney	University of Ulster, UK
Christian Fischer Pedersen	Aarhus University, Denmark
Christophe Trefois	University of Luxembourg, Luxembourg
Dana Petcu	West University of Timisoara, Romania
David Guralnick	International E-Learning Association, USA
David Šafránek	Masaryk University, Czech Republic
Dejan Gjorgjevikj	Ss. Cyril and Methodius University, North Macedonia
Dejan Spasov	Ss. Cyril and Methodius University, North Macedonia
Denis Trcek	University of Ljubljana, Slovenia
Dilip Patel	London South Bank University, UK
Dimitar Trajanov	Ss. Cyril and Methodius University, North Macedonia
Eftim Zdravevski	Ss. Cyril and Methodius University, North Macedonia
Elena Vlahu-Gjorgievska	University of Wollongong, Australia
Emmanuel Conchon	University of Limoges, France
Florin Pop	University Politehnica of Bucharest, Romania
Francesc Burrull	Universidad Politecnica de Cartagena, Spain

Fu-Shiung Hsieh	Chaoyang University of Technology, Taiwan
Georgina Mirceva	Ss. Cyril and Methodius University, North Macedonia
Gjorgji Madjarov	Ss. Cyril and Methodius University, North Macedonia
Goce Armenski	Ss. Cyril and Methodius University, North Macedonia
Goran Velinov	Ss. Cyril and Methodius University, North Macedonia
Hieu Trung Huynh	Industrial University of Ho Chi Minh City, Vietnam
Hrachya Astsatryan	National Academy of Sciences of Armenia, Armenia
Hristijan Gjoreski	Ss. Cyril and Methodius University, North Macedonia
Hristina Mihajloska	Ss. Cyril and Methodius University, North Macedonia
Igor Ljubi	University of Zagreb, Croatia
Ilche Georgievski	University of Stuttgart, Germany
Irina Mocanu	University Politehnica of Bucharest, Romania
Ivan Chorbev	Ss. Cyril and Methodius University, North Macedonia
Ivan Kitanovski	Ss. Cyril and Methodius University, North Macedonia
Ivica Dimitrovski	Ss. Cyril and Methodius University, North Macedonia
Jatinderkumar Saini	Narmada College of Computer Application, India
John Gialelis	University of Patras, Greece
Josep Silva	Universitat Politècnica de València, Spain
Jugoslav Achkoski	Military Academy "General Mihailo Apostolski", North Macedonia
Kalinka Kaloyanova	University of Sofia, Bulgaria
Kalinka Regina Castelo Branco	University of São Paulo, Brazil
Kaori Fujinami	Tokyo University of Agriculture and Technology, Japan
Katarina Trojacanec	Ss. Cyril and Methodius University, North Macedonia
Katerina Zdravkova	Ss. Cyril and Methodius University, North Macedonia
Kosta Mitreski	Ss. Cyril and Methodius University, North Macedonia
L. T. Chitkushev	Boston University, USA

Natasha Ilievska	Ss. Cyril and Methodius University, North Macedonia
Nevena Ackovska	Ss. Cyril and Methodius University, North Macedonia
Neville Calleja	University of Malta, Malta
Nikola Simidjievski	University of Cambridge, UK
Pance Panov	Jozef Stefan Institute, Slovenia
Panche Ribarski	Ss. Cyril and Methodius University, North Macedonia
Pece Mitrevski	University "St. Kliment Ohridski" - Bitola, North Macedonia
Petre Lameski	Ss. Cyril and Methodius University, North Macedonia
Riste Stojanov	Ss. Cyril and Methodius University, North Macedonia
Robertas Damasevicius	Silesian University of Technology, Poland
Rodica Potolea	Technical University of Cluj-Napoca, Romania
Rossitza Goleva	New Bulgarian University, Bulgaria
Sanja Lazarova-Molnar	University of Southern Denmark, Denmark
Sasho Gramatikov	Ss. Cyril and Methodius University, North Macedonia
Sasko Ristov	University of Innsbruck, Austria
Saso Dzeroski	Jozef Stefan Institute, Slovenia
Sergio Ilarri	University of Zaragoza, Spain
Shushma Patel	De Montfort University, UK
Shuxiang Xu	University of Tasmania, Australia
Simona Samardjiska	Radboud University, The Netherlands
Slobodan Bojanic	Universidad Politécnica de Madrid, Spain
Slobodan Kalajdziski	Ss. Cyril and Methodius University, North Macedonia
Smile Markovski	Ss. Cyril and Methodius University, North Macedonia
Smilka Janeska-Sarkanjac	Ss. Cyril and Methodius University, North Macedonia
Snezana Savoska	University "St. Kliment Ohridski" - Bitola, North Macedonia
Sonja Gievska	Ss. Cyril and Methodius University, North Macedonia
Stanimir Stoyanov	University of Plovdiv "Paisii Hilendarski", Bulgaria
Suliman Mohamed Fati	Prince Sultan University, Saudi Arabia
Suzana Loshkovska	Ss. Cyril and Methodius University, North Macedonia

Syed Ahsan	Technische Universität Graz, Austria
Tome Eftimov	Jozef Stefan Institute, Slovenia
Ustijana Rechkoska-Shikoska	UIST, North Macedonia
Vacius Jusas	Kaunas University of Technology, Lithuania
Verica Bakeva	Ss. Cyril and Methodius University, North Macedonia
Vesna Dimitrova	Ss. Cyril and Methodius University, North Macedonia
Vladimír Siládi	Matej Bel University, Slovakia
Vladimir Trajkovik	Ss. Cyril and Methodius University, North Macedonia
Wuyi Yue	Konan University, Japan
Xiangyan Zeng	Fort Valley State University, USA
Yoram Haddad	Jerusalem College of Technology, Israel
Zahid Akhtar	University of Udine, Italy
Zaneta Popeska	Ss. Cyril and Methodius University, North Macedonia
Zoran Zdravev	Goce Delcev University of Stip, North Macedonia

Contents

Keynote

A Gentle Introduction to Zero-Knowledge

Andrej Bogdanov$^{(\boxtimes)}$

Department of Computer Science and Engineering and Institute of Theoretical
Computer Science and Communications, The Chinese University of Hong Kong,
Shatin, Hong Kong
andrejb@cse.cuhk.edu.hk

Abstract. Zero-knowledge proofs, invented in the 1980s, allow one to
certify the validity of a statement without revealing why it is true. I
illustrate their relevance to secure identification, blockchain privacy, and
electronic voting. I also discuss their generality and composability.

Interactions in society are based on trust. Even in the early days of the internet it was recognized that trust cannot be replicated online. Lack of trust in settings such as information dissemination are a considerable challenge of our times. For situations with a clearly defined objective (e.g. identification, voting, poker), however, in the 1980s cryptographers discovered that trusted parties (e.g. law enforcement, electoral commissions, the casino) can in theory be eliminated and replaced by *interactive protocols* that guarantee an indistinguishable outcome even if some of the participants behave maliciously.

This talk is about a key conceptual insight called a "zero-knowledge proof" [GMR89] that is used to enforce honest behavior in such protocols without leaking any unintended information. While zero-knowledge proofs have been available for over three decades now recent tendencies to store information in shared databases such as blockchains is making them much more relevant to practice.

As an example, imagine the digital equivalent of *identification* to a online entity with whom one wants to carry out a digital transaction. In a physical situation, one can use a certified document such as a government-issued ID card or passport for this purpose. Making this information public, say by posting it on a blockchain widely reveals private information such as date of birth and place of residence. A preferable alternative is to reveal only the necessary information, namely to *prove* a statement like "I am the holder of identity number 32571 *and* the holder of identity number 32571 has at least 7 dollars in their bank account" without revealing anything additional. Zero-knowledge proofs are precisely that: Proofs of theorems that reveal nothing beyond the truth of the theorem in question.

This article was written while visiting the Simons Institute for the Theory of Computing
at UC Berkeley.

L. Antovski and G. Armenski (Eds.): ICT Innovations 2021, CCIS 1521, pp. 3–10, 2022.
https://doi.org/10.1007/978-3-031-04206-5_1

An Example of a Zero-Knowledge Proof

The key features of a zero-knowledge proof are captured in the following somewhat abstract but insightful example of Goldreich, Micali, and Wigderson [GMW91].[1] Consider the following two matrices of bits:

$$A = \begin{matrix} 0\,1\,1\,0 \\ 1\,0\,1\,0 \\ 1\,0\,1\,1 \\ 1\,1\,0\,1 \end{matrix} \qquad B = \begin{matrix} 1\,1\,1\,0 \\ 1\,0\,0\,1 \\ 0\,1\,0\,1 \\ 0\,1\,1\,1 \end{matrix} \qquad (1)$$

A theorem about these two matrices is that they are *isomorphic*: The rows and columns of A can be rearranged to obtain B.

The common way to prove this theorem is to describe how the rows and columns of A should be rearranged, namely to describe permutations π and σ of the rows and columns of A, respectively, such that $\pi A \sigma = B$:

$$A = \begin{matrix} 0\,1\,1\,0 \\ 1\,0\,1\,0 \\ 1\,0\,1\,1 \\ 1\,1\,0\,1 \end{matrix} \Big\}\, \pi \qquad \pi A = \begin{matrix} 1\,1\,0\,1 \\ 0\,1\,1\,0 \\ 1\,0\,1\,0 \\ 1\,0\,1\,1 \end{matrix} \qquad \overset{\sigma}{\longleftrightarrow} \qquad \pi A \sigma = B = \begin{matrix} 1\,1\,1\,0 \\ 1\,0\,0\,1 \\ 0\,1\,0\,1 \\ 0\,1\,1\,1 \end{matrix}$$

This proof is not zero knowledge because not only does it show that A and B are isomorphic but it also reveals what the isomorphism (π, σ) is, namely *how* to go from A to B. If (π, σ) represents a "secret key", once this proof has been written down the key is no longer secret.

To understand how a zero-knowledge proof for this theorem looks like it is helpful to revisit the requirements of a proof. Any logic for proving theorems has to satisfy two criteria. It has to be *complete*, meaning that it should be in principle able to prove all true statements of the type "A and B are isomorphic". It also has to be *sound*, namely if A and B happen not to be isomorphic, no purported proof should pass verification. This is indeed the case for the proof system that was just described: If A and B are not isomorphic, no matter which "proof" (π, σ) is proposed, it is easy to check that $\pi A \sigma$ does not equal B.

The zero-knowledge proof system for graph isomorphism will differ from a typical mathematical proof in two respects: It will be *interactive* and it will be *randomized*. The proof is no longer a written string of symbols but a game that involves two parties: A *Prover* and a *Verifier*. The input matrices A and B are known to both. The logical requirements are:

- **Completeness:** If A and B are isomorphic, an honest Prover who knows π and σ can convince the Verifier that they are.
- **Soundness:** If A and B are not isomorphic, no Prover, honest or dishonest, can convince the Verifier that they are.

---

[1] A minor technical difference is that the example of Goldreich et al. concerns isomorphism of graphs while ours is about isomorphism of matrices, i.e., bipartite graphs.

To describe how the proof works, we need to describe the "code" of the honest Prover and of the Verifier. In the classical proof system, the Prover writes down π and σ and the verifier simply checks whether $\pi A \sigma$ equals B as in the above example.

Common input: Matrices A and B

Honest Prover's additional input: Permutations π and σ such that $\pi A \sigma = B$.

1 **Honest Prover:** Choose random permutations π', σ'.
   Send the matrix $C = \pi' B \sigma'$ to the Verifier.
2 **Verifier:** Choose a random bit $b \in \{0, 1\}$.
   Send b to the Prover.
3 **Honest Prover:**
   If $b = 1$ send π' and σ' to the Verifier.
   If $b = 0$ send $\pi'\pi$ and $\sigma\sigma'$ to the Verifier.
4 **Verifier:** Upon receiving ρ and τ,
   If $b = 0$ and $\rho A \tau = C$, accept the proof as valid.
   If $b = 1$ and $\rho B \tau = C$, accept the proof as valid.
   Otherwise, reject the proof.

Here, $\pi'\pi$ is the composition of permutations π' and π, namely the permutation obtained by permuting the rows using π and then permuting them again using π'. Similarly, $\sigma\sigma'$ is the row permutation obtained by applying σ followed by σ'. Let us show a sample run of this protocol on the same example. For example, in step 1 the Honest Prover could have done the following:

$$B = \begin{matrix} 1 & 1 & 1 & 0 \\ 1 & 0 & 0 & 1 \\ 0 & 1 & 0 & 1 \\ 0 & 1 & 1 & 1 \end{matrix} \quad \pi'B = \begin{matrix} 1 & 0 & 0 & 1 \\ 0 & 1 & 0 & 1 \\ 1 & 1 & 1 & 0 \\ 0 & 1 & 1 & 1 \end{matrix} \quad \pi'B\sigma' = C = \begin{matrix} 1 & 1 & 0 & 0 \\ 1 & 0 & 1 & 0 \\ 0 & 1 & 1 & 1 \\ 1 & 0 & 1 & 1 \end{matrix}$$

If in step 2 the Verifier chooses $b = 0$, the Honest Prover will compute the permutations

$$\pi' \circ \pi = \rho \qquad \sigma \circ \sigma' = \tau$$

and the verifier will calculate

$$A = \begin{matrix} 0 & 1 & 1 & 0 \\ 1 & 0 & 1 & 0 \\ 1 & 0 & 1 & 1 \\ 1 & 1 & 0 & 1 \end{matrix} \quad \rho A = \begin{matrix} 0 & 1 & 1 & 0 \\ 1 & 0 & 1 & 0 \\ 1 & 1 & 0 & 1 \\ 1 & 0 & 1 & 1 \end{matrix} \quad \rho A \tau = \begin{matrix} 1 & 1 & 0 & 0 \\ 1 & 0 & 1 & 0 \\ 0 & 1 & 1 & 1 \\ 1 & 0 & 1 & 1 \end{matrix}$$

which yields the matrix C and therefore acceptance of the proof. In general, if $b = 0$ then $\rho A \tau = \pi'\pi A \sigma\sigma' = \pi' B \sigma' = C$, and if $b = 1$ then $\rho B \tau = \pi' B \sigma' = C$, so the Verifier always accepts interactions with the Honest Prover.

What happens if A and B are *not* isomorphic matrices? In this case there is no honest prover as π and σ do not even exist. What we will argue is that no matter how the Prover works, regardless of the messages sent in steps 1 and 3 the Verifier is unlikely to accept. Suppose that the Verifier accepted the proofs both in the cases $b = 0$ and $b = 1$. Then it must be that both A is isomorphic to C ($\rho_0 A \tau_0 = C$ for some ρ_0, τ_0) *and* B is isomorphic to C ($\rho_1 B \tau_1 = C$ for some ρ_1, τ_1), so A must have been isomorphic to B ($\rho_1^{-1} \rho_0 A \tau_0 \tau_1^{-1} = B$)!

In conclusion, no matter what a cheating prover does, the Verifier can always pick a suitable value of b that detects this cheating behavior. As b is a random bit we can conclude that Verifier detects cheating with probability at least $1/2$. In this example the probability is in fact exactly $1/2$ in the sense that there exists a cheating Prover that can make the Verifier accept half the time. Unlike a classical proof which is either correct or incorrect, in an interactive proof there is some chance that an incorrect proof is accepted as valid. In this example the probability of such an event happening is $1/2$. This *soundness error* can be reduced to a very small number like 2^{-50} by repeating the interaction 50 times.

So far the proof system that we have described is inferior to the classical way of proving isomorphism of matrices: It requires interaction and it has a soundness error (in addition to being more complicated). Now we come to the crucial question: What did the Verifier learn after interacting with the Honest Prover?

To answer this question we need to look at the interaction from the perspective of the Verifier. In step 1, the Verifier saw the matrix C which was obtained by applying a random isomorphism (π', σ') to B. In step 3 it observed either this random isomorphism (π', σ') if it chose $b = 1$, or another isomorphism $(\pi'\pi, \sigma'\sigma)$ if it chose $b = 0$. But no matter what π and σ are, the isomorphism $(\pi'\pi, \sigma'\sigma)$ is itself a random isomorphism!

Thus, the transcript of interaction observed by the verifier comes from the following distribution: First, sample a random bit b, then sample random permutations ρ and τ, then set $C = \rho A \tau$ if $b = 0$ and $C = \rho B \tau$ if $b = 1$. Sampling the triplet $(C, b, (\rho, \tau))$ now requires *no input and no interaction*, yet is identically distributed to the one that the Verifier observed in the actual interaction with the Honest Prover. Thus we can sensibly conclude that the Verifier learned nothing from this interaction—nothing beyond the fact that he is interacting with an Honest Prover and therefore A and B are indeed isomorphic matrices.

We thus come to the key feature of zero-knowledge proofs: Whatever knowledge the Verifier gained after interacting with the Honest Prover, he could have generated on his own. This perspective allows one to argue rigorously about security of interactions with respect to learning unintended information—without ever having to define what knowledge is!

Applications and Variants

The simulation paradigm of zero-knowledge comes up virtually every time one has to argue about the security of a cryptographic protocol. I will highlight three representative examples.

Electronic Voting. An election between two candidates can be modeled as a computation of the function $x_1 + x_2 + \cdots + x_n$, where $x_i \in \{0,1\}$ represents the i-th voter's choice—0 for Alice, 1 for Bob. A common requirement is that the ballot should be secret. In electronic voting systems, this is commonly enforced by encrypting the votes. It is useful that the underlying encryption schemes be *additively homomorphic* so that the votes can be tallied before they are decrypted (e.g., [DGS03]). One example of an additively homomorphic encryption scheme is El Gamal encryption, in which a vote for candidate x has the format $(g^r, PK^r g^x)$. Exponentiation is taken modulo some cyclic group with generator g; PK and r describe the public key and internal randomness of the encryption. In a naive implementation a candidate can submit votes like $x = 100$ or $x = -17$, which would be the analogue of voting for Bob 100 times or taking 17 votes away from him. Zero-knowledge proofs are a mechanism for protecting against such attacks: In addition to submitting the actual encrypted vote $Enc(x; r) = (g^r, PK^r g^x)$ the voter also provides a zero knowledge proof of the claim "$x = 0$ or $x = 1$."

Identification. The username/password mechanism is the most common method for identification to an online service. In such a scenario the user acts as a prover and the service is the verifier. A malicious verifier that intercepts the password (e.g., a phishing website) can then impersonate the user in future interactions with the service. In fact, any non-interactive identification mechanism such as passwords is susceptible to such *replay attacks*.

Zero-knowledge protocols are closely related to interactive identification. In the matrix isomorphism example, we can view the matrices A and B as a "public key" and the permutations π and σ as the user's associated "secret key". The objective of the interaction is however different: While the task of the Honest Prover was to convince the Verifier that the matrices A and B *are* isomorphic, the user's objective in an identification protocol is to convince the service that she *knows* her secret key, namely the isomorphism π, σ for which $\pi A \sigma$ equals B. This is a more stringent requirement because if a user knows the isomorphism, then in particular an isomorphism must exist. Although the converse is not in general true (or even sensible to state), the proof system that we described satisfies the more stringent requirement of a *zero-knowledge proof of knowledge* [BG93].

A secure identification protocol has to satisfy the additional requirement that it is in general hard to extract information about the secret key from the public key. While this is not known to hold for matrix isomorphisms, it is believed to be true for other problems such as taking discrete logarithms in suitable groups [Sch90].

Blockchain Anonymity. One distinguishing feature of blockchains in contrast to centralized ledgers is anonymity: As users are identified by their public keys only, their identity is unknown. This is especially important as all transactions are publicly visible. Suppose, however, that you see the following transaction record:

User 57269 buys product 34130 for 10 bitcoins.

Ten bitcoins is a large amount, immediately narrowing down the possible users that have ID 57269 to a small number of possibility. If, in addition, you knew that this transaction was posted by someone from Skopje, the number of suspects would drop to a handful.

A greater level of anonymity could be achieved by hiding the transaction amount:

User 57269 buys product 34130 for "40762" bitcoins.

The number 40762 no longer represents the actual amount, but is a *commitment* to such a number. You can think of it as the output of some hard-to-invert "hash function" H applied to the actual bitcoin amount 10 followed by some random bits R. To prove that this transaction is valid, i.e. to ensure that user 57629 has enough bitcoins for the purchase, the transaction should be accompanied by a proof of the following claim:

I know the bitcoin amount Z and random bits R such that $H(Z, R) = 40762$ and user X has a balance of at least Z bitcoins.

If this proof is effected in zero-knowledge then the seller of product 34130 is confident that the transaction is valid, while no public information is revealed about the finances of user 57269 or the price of product 34130. Some cryptocurrencies like zcash provide such enhanced anonymity thanks to zero-knowledge [PHGR16].

As blockchain transactions are written records, the corresponding zero-knowledge proof must be *non-interactive*. This can be accomplished at the cost of some additional assumptions [BDSMP91, FLS99].

Generality and Composability

Claims of type "matrices A and B are isomorphic" are of restricted interest. How general are the statements that admit zero-knowledge proofs? Soon after the discovery of zero-knowledge it was shown that any written proof (in a sufficiently powerful logical system) can be converted into a zero-knowledge proof [GMW91], thus *everything provable is provable in zero-knowledge*.

To give a sense of this power, suppose that you are given two logical statements A, B. If you have a proof of one of them (but not the other), then you can claim the statement "A or B". If you have a proof of both, then you can claim the statement "A and B".

Now suppose you have a zero-knowledge proof of statement A and another zero-knowledge proof of statement B. Can you prove "A and B" in zero-knowledge? If the Honest Prover and the Verifier run the protocol for A followed by the protocol for B, the resulting proof would indeed be zero-knowledge. Intuitively, if you know that A and B are true but nothing else then you know that each of them is true individually, but nothing else.

The same type of reasoning fails for statements of the form "A or B". Suppose A was true and B was false. If I proved "A or B" by showing you the proof for A you would be learning additional information, namely which of the two statements is true. The resulting proof would not be zero-knowledge.

There are, however, more intricate composition methods for proving the OR of two statements that hide the identity of the true statement [CDS94, AOS02]. As both AND and OR can be composed in zero-knowledge, it is in principle possible to take any monotone predicate P of some atomic claims $A_1, \ldots, A_n$ that are provable in zero-knowledge, represent it logically using ANDs and ORs of the A_is, and obtain a zero-knowledge proof of P. In general, however, the amount of interaction in the zero-knowledge proof will depend on the complexity of the representation.

For concreteness, consider the following type of predicate P. There is some underlying directed graph G with two distinguished nodes s and t. Each edge in this graph is associated with an atomic claim. For example, nodes could represent financial entities and the claims on the edges could represent transfers between them recorded in some ledger. An auditor which controls some of the nodes in G may want to prove that he can observe *all* cash flows from s to t without revealing which nodes she controls. It turns out that the computational model in which P is represented substantially affects the complexity of the representation. In a recent work that I was involved in [AAB+21] we worked out a zero-knowledge proof system that is particularly efficient for predicates of this type.

Conclusion and Further Reading

The technological significance of zero-knowledge proofs will undoubtedly rise as information is increasingly shared and secure multiparty computation becomes more ubiquitous. I believe, however, that the main value of zero-knowledge is not technical but conceptual: The definition allows for rigorous reasoning about unintended information disclosure in a wide variety of settings. It is important, however, to also be aware of their limitations: While the guarantee of revealing no information beyond the truth of the statement is the strongest possible for a complete and sound proof system, it is sometimes inadequate. In such scenarios mitigating privacy loss may necessitate some amount of lying [DMNS06].

A rigorous treatment of zero-knowledge proofs with extensive references is given in Chap. 4 in volume 1 of Goldreich's book *Foundations of Cryptography* [Gol06] (see also Appendix C in volume 2 [Gol09]).

References

[AAB+21] Abe, M., Ambrona, M., Bogdanov, A., Okhubo, M., Rosen, A.: Acyclicity programming for sigma-protocols. In: TCC 2021: Theory of Cryptography (2021, to appear)

[AOS02] Abe, M., Ohkubo, M., Suzuki, K.: 1-out-of-n signatures from a variety of keys. In: Zheng, Y. (ed.) ASIACRYPT 2002. LNCS, vol. 2501, pp. 415–432. Springer, Heidelberg (2002). https://doi.org/10.1007/3-540-36178-2_26

[BDSMP91] Blum, M., De Santis, A., Micali, S., Persiano, G.: Noninteractive zero-knowledge. SIAM J. Comput. **20**(6), 1084–1118 (1991)

[BG93] Bellare, M., Goldreich, O.: On defining proofs of knowledge. In: Brickell, E.F. (ed.) CRYPTO 1992. LNCS, vol. 740, pp. 390–420. Springer, Heidelberg (1993). https://doi.org/10.1007/3-540-48071-4_28

[CDS94] Cramer, R., Damgård, I., Schoenmakers, B.: Proofs of partial knowledge and simplified design of witness hiding protocols. In: Desmedt, Y.G. (ed.) CRYPTO 1994. LNCS, vol. 839, pp. 174–187. Springer, Heidelberg (1994). https://doi.org/10.1007/3-540-48658-5_19

[DGS03] Damgård, I., Groth, J., Salomonsen, G.: The theory and implementation of an electronic voting system. In: Gritzalis, D.A. (ed.) Secure Electronic Voting, pp. 77–99. Springer, Boston (2003). https://doi.org/10.1007/978-1-4615-0239-5_6

[DMNS06] Dwork, C., McSherry, F., Nissim, K., Smith, A.: Calibrating noise to sensitivity in private data analysis. In: Halevi, S., Rabin, T. (eds.) TCC 2006. LNCS, vol. 3876, pp. 265–284. Springer, Heidelberg (2006). https://doi.org/10.1007/11681878_14

[FLS99] Feige, U., Lapidot, D., Shamir, A.: Multiple noninteractive zero knowledge proofs under general assumptions. SIAM J. Comput. **29**(1), 1–28 (1999)

[GMR89] Goldwasser, S., Micali, S., Rackoff, C.: The knowledge complexity of interactive proof systems. SIAM J. Comput. **18**(1), 186–208 (1989)

[GMW91] Goldreich, O., Micali, S., Wigderson, A.: Proofs that yield nothing but their validity or all languages in np have zero-knowledge proof systems. J. ACM **38**(3), 690–728 (1991)

[Gol06] Goldreich, O.: Foundations of Cryptography: Volume 1, Basic Tools. Cambridge University Press, Cambridge (2006)

[Gol09] Goldreich, O.: Foundations of Cryptography: Volume 2, Basic Applications. Cambridge University Press, Cambridge (2009)

[PHGR16] Parno, B., Howell, J., Gentry, C., Raykova, M.: Pinocchio: nearly practical verifiable computation. Commun. ACM **59**(2), 103–112 (2016)

[Sch90] Schnorr, C.P.: Efficient identification and signatures for smart cards. In: Brassard, G. (ed.) CRYPTO 1989. LNCS, vol. 435, pp. 239–252. Springer, New York (1990). https://doi.org/10.1007/0-387-34805-0_22

Deep Learning and AI

Automated Assessment of Lower and Higher-Order Thinking Skills Using Artificial Intelligence Methods

Emil Hadzhikolev, Stanka Hadzhikoleva(✉), Kostadin Yotov, and Maria Borisova

Faculty of Mathematics and Informatics, University of Plovdiv "Paisii Hilendarski",
236 Bulgaria Bul., Plovdiv, Bulgaria
{hadjikolev,stankah,kostadin_yotov}@uni-plovdiv.bg,
mimi880503@gmail.com

Abstract. The complex assessment of the higher-order thinking skills is a challenge that every teacher has faced. Often in the education process, teachers test and assess students' knowledge and skills in different ways – through tests, assignments, case studies, answers to questions, essays, term papers, etc. The final grade is formed by a formula, as a function of the grades obtained. In some cases, teachers find inexplicable gaps between the grades received by a student. For example – a student receives a low grade for knowledge of basic concepts and facts, and a high one for solving a complex problem that requires excellent knowledge of these basic concepts and facts. Striving to be as objective as possible and fair at the same time, many teachers in such a situation make a subjective assessment, based on their personal opinion about the student and their teaching experience. The article proposes an approach for automated complex assessment, using methods of artificial intelligence. Experiments have been conducted using Neural Networks, SVM and Linear Regression algorithms. The Orange application was used to conduct the study.

Keywords: Student assessment · Formative assessment · Assessment of HOTS and LOTS · Assessment using artificial intelligence methods

1 Introduction

The process of testing and assessing the knowledge and skills of students is no less important than the process of learning itself. The objective and fair assessment shows the students their weaknesses and shortcomings, tells them in which direction to improve, gives them a basis for comparison with other students, and in the best case – confidence that they are doing well. On the other hand, through the assessments the teacher can find knowledge and skills that are difficult to learn, identify potential areas for creative development of the students, draw conclusions about the applicability of the methods and tools used for teaching, as well as the degree of mastery of the curriculum.

Assessment Has Many Aspects. It can be considered as a process of comparing the results achieved in the educational work with the previously set goals. *The verification*

© Springer Nature Switzerland AG 2022
L. Antovski and G. Armenski (Eds.): ICT Innovations 2021, CCIS 1521, pp. 13–25, 2022.
https://doi.org/10.1007/978-3-031-04206-5_2

and assessment are also part of the teacher's control in the teaching process, which allows through various methods, forms and means to establish and check in advance the current or final state of the volume and quality of the students' knowledge, skills and competencies [1]. Many educators believe that *the assessment the teacher makes for the knowledge and skills of the student is a strong motivating factor for students when it is objective, positive and stimulating for them*. Then it can become a means of developing positive motives for learning.

It is indisputable that in order for the assessment to be objective, reliable and valid, *significant preliminary work is needed to develop and improve a system of forms, methods and means for verification and assessment*. The learning process uses a wide variety of assessment forms: tests, essays, assignments, reports, term papers, case studies, presentations etc. The final grade in the general case is formed as a function of the assessment components, as each assessment component participates in it with different weight.

Many teachers have found it difficult to form a final grade in certain problematic situations. There are different types of problematic situations. In some cases, the students did very well in the practical tasks during the training period but failed the theoretical exam. In other cases, they know the theory very well, but receive low grades for practical assignments. The difference in these assessments is due to *the different degree of mastery of lower-order thinking skills (LOTS) and of higher-order thinking skills (HOTS)*.

The article presents *an experiment for automated formation of a final assessment of students, based on different assessment components, assessing HOTS and LOTS*. Three methods of artificial intelligence (AI) for machine learning have been used - Neural Networks (NNs), SVM, and Linear Regression. They first use real data for grades set by a teacher and their corresponding final grades to create a mathematical model. Then they predict the students' final grades as a function of their current component grades. The student-grade datasets we had were not enough to create a good neural network. For this reason, we used the GARP method, created by us, to randomly generate additional data samples to increase the neural network training dataset. With this, we achieved the creation of a neural network that works with greater accuracy. The Orange application was used to perform the experiment [2]. It is open source and supports a variety of machine learning and data visualization tools that can be used to construct various workflows for data analysis and forecasting. The experiment we conducted also aims to show that *teachers can easily use artificial intelligence methods to support their work related to the analysis and assessment of student knowledge and skills*.

2 Assessment of Higher-Order Thinking Skills

Many training systems and models use the *Bloom's Taxonomy* [3]. It is a hierarchy of thinking skills at 6 levels, in which the higher cognitive levels include and upgrade all cognitive skills form the lower levels. According to it, the cognitive sphere has six main levels, arranged on the principle of "simple to complex" – Knowledge, Comprehension, Application, Analysis, Synthesis, Evaluation. Each of these levels defines a number of cognitive goals and activities through which goals can be achieved.

Bloom's theory is based on the idea that *goals and learning outcomes are not the same*. For example, memorizing scientific facts, no matter how important, is at a lower level than the ability to analyze or assess. To assess a given process, the student must analyze it first. To apply a concept, he must first understand it. Thus, each of Bloom's levels builds on the previous one. In the process of learning a subject, students move sequentially from the lowest to the highest levels of knowledge and skills.

Table 1. Bloom's Taxonomy – cognitive areas, goals and activities

Cognitive area	General goals	Keywords, activities
Knowledge	Knows the specific facts, knows the general terminology, basic concepts and functions	Describes, lists, points out, reproduces, formulates, names, chooses, emphasizes
Comprehension	Understands facts, principles, explains methods and procedures, assesses consequences	Transforms, defends, distinguishes, assesses, explains, expands, summarizes, gives examples, paraphrases, predicts, retells, defines, discusses, compares
Application	Applies concepts and principles in new situations, demonstrates proper use of new method or procedure. Solves problems/tasks with the help of acquired knowledge and skills	Uses, proves, discovers, solves, changes, develops
Analysis	Recognizes implicitly formulated ideas and assumptions. Recognizes logical paradoxes in reasoning. Distinguishes facts from conclusions. Assesses the applicability of the data. Analyzes the structure of a plan, project. Makes a breakdown of the material into its components	Divides into components, presents graphically, distinguishes, differentiates, defines, illustrates, draws conclusions, summarizes, indicates, connects, divides
Synthesis	Develops new ideas, organizes the presentation in a written text. He has a well-developed and organized speech. He writes stories. Develops a project plan. Integrates what has been learned in other areas into a problem-solving plan	Categorizes, combines, collects, complies, creates, invents, designs, explains, generates, modifies, rearranges, reconstructs, recognizes, revises
Evaluation	Assesses the logical coherence of a written material, the adequacy of the data supporting the conclusions. Assesses the value of the work using external quality standards. Assesses the value of the work using internal criteria	Assesses, categorizes, compares, draws conclusions, criticizes, describes, explains, differentiates, proves, interprets, connects, summarizes, points out, defends

The thinking skills of the three lower levels of the taxonomy – Knowledge, Comprehension, and Application – are referred to as *lower-order thinking skills*, and those of the higher levels – Analysis, Synthesis, and Evaluation, as *higher-order thinking skills*. Table 1 shows the main cognitive areas of Bloom's Taxonomy and the corresponding appropriate activities for their development and improvement [3].

If the students only memorize and reproduce information, understand and explain concepts and ideas only in general, and if they apply information and rules only in familiar situations, they will not acquire HOTS. To achieve this higher level of thinking activity, students need to perform other types of activities – analyzing information to explore knowledge and relationships, assessing decisions or ways of doing things, creating new ideas, products and perspectives, etc. *The teachers' knowledge of HOTS and the approaches to their teaching and learning are extremely important for the successful conduct of the education* [4].

We believe that systems for testing students' knowledge and skills should be designed in a way that differentiates the assessment of different types of higher- and lower-order thinking skills. For this purpose, it is necessary to use *different forms and methods of testing and evaluation to assess different cognitive skills and different aspects of knowledge*. In his book "How to Assess Higher-Order Thinking Skills in Your Classroom", Brookhar examines in great detail the different approaches to assessing HOTS. It can be used as a source of ideas for teachers [5]. An interesting approach to assessing HOTS is described in [6]. The authors present student achievements through numerical matrices, with each row describing Bloom's cognitive level and each column representing a specific task. Each value in the matrix indicates the extent to which the student has been able to apply the skills of the respective cognitive level, determined by the order of the matrix, on the task determined by the column of the matrix. The authors study matrix behavior and thus draw conclusions about the students' thinking skills. Another model proposed by us for multicomponent fuzzy evaluation, with a focus on the assessment of HOTS is described in [7, 8].

Modern Learning Management Systems provide a wide variety of assessment tools, through quizzes, assignments, games, integration of assessment plug-ins, etc. Collaboration tools (such as blogs, wikis, workshops, databases, and glossaries) also provide an opportunity to assess teamwork and soft skills, taking into account the individual performance of each student [9]. In many cases, the teacher can review their exam resources, judge what knowledge and skills according to Bloom are being assessed, and classify them as HOTS or LOTS assessment tasks. He can then track and analyze the students' achievements [10]. Other teachers prefer to develop assessment tools following established methods [11]. Undoubtedly, this allows more precise development and adjustment of the assessment components and their focus on HOTS.

The assessment of HOTS must take into account the possibilities of the students for a partial solution to the set task. It is appropriate *to define a number of criteria that assess the various components of the overall solution of the task* [12]. Here we should note that *the periodic conducting of quality research of education improves the educational process and increases the confidence in the grades* [13].

Assessing student knowledge and skills and forming a fair final grade is in some cases a difficult task, even for experienced teachers. For example, if the student has received a high assessment on a complex task requiring HOTS, and at the same time has a low grade on a test requiring LOTS, which assesses the acquisition of minimum knowledge to solve the complex task. For each such case, the teacher will have to consider the student's achievements so far and his other grades in the discipline before forming a final grade. The accumulation of information on such specific cases of student grades can be used to perform automated assessments using artificial intelligence.

3 Use of AI Methods for Formative Assessments

Artificial intelligence has quickly entered our daily lives. The cheaper technology and the many free and open-source software tools, frameworks and libraries allow any programmer to easily use the latest advances in science.

Three methods of artificial intelligence were used in this study – Neural Networks, SVM and Linear Regression [14, 15]. The experiments were performed using the Orange application. It is a free and open-source application for data visualization, analysis and data predictions. It supports various data processing options that can be implemented through visual programming or Python scripts. The application has a rich set of tools for data analysis, machine learning, bioinformatics, text extraction, etc. It can be used both by people with no experience in programming and using prediction methods, as well as by professionals to quickly solve data prediction and classification tasks.

Orange application has its strengths and weaknesses. It supports excellent visualization of various processes, enables machine learning with "one click" and easy redirection of data flow to different modules. On the other hand, there are weaknesses in some of the tools. For example – when working with neural networks, it mostly supports algorithms that require fewer computer resources and allow the use of more neurons and iterations. Therefore, fine-tuning the networks can be a problem in some cases, which also means difficulties in achieving greater accuracy.

3.1 Setting up the Experiment

The experiment was conducted on the assessments made in the process of teaching two groups of students, 59 and 77 people, respectively. They study in two different specialties in the professional field "Informatics and Computer Science" at Plovdiv University "Paisii Hilendarski". The subject of the research are the results of the assessment in the discipline "Internet programming with PHP and MySQL".

Four types of assessment components have been constructed to assess the learners' knowledge and skills, as follows:

- **T_LOTS** – assesses basic lower-order thinking skills focused on the theoretical material being studied – **acquired knowledge, the ability to understand and apply** the studied theoretical material, etc. It is conducted through a test, where the questions require **distinguishing, listing and comparing objects; explaining and giving examples of concepts; reproduction, explanation and use of concepts, etc**. The test score is an integer in the interval [0, 30], we denote it by x_1.

- **P_LOTS** – assesses practical knowledge of lower-order. The abilities for **understanding and application** of the syntactic constructions, libraries, etc. studied in the discipline are assessed. Test items include questions about **predicting the result of the execution of a short program code, code modification, error detection, application of examples, etc.** The grade is an integer in the interval [0, 15], we denote it by x_2.
- **T_HOTS** – assesses higher-order theoretical thinking skills. The students demonstrate their abilities for **analysis, synthesis and assessment**. The questions are related to **explanation, interpretation, graphical presentation and modification of program code; comparison and critical analysis of decisions; design of software components; summarization, proving, drawing conclusions, assessments, etc.** The grade is an integer in the interval [0, 15], we denote it by x_3.
- **P_HOTS** – assesses the practical potential of the students to solve relatively new tasks, to **design and develop** web applications. The grade is an integer in the interval [2, 6], we denote it by x_4.

The components x_1, x_2 and x_3 are related to the material taught in lectures. The components x_4 assesses the practical work of students on exercises. Thus, each student has four grades, and based on them, the teacher forms a final grade. In the experiment, all final grades were formed by only one teacher.

3.2 Use of Neural Networks

Neural networks are based on the idea of a mathematical interpretation of a biological neuron. Modeled with the means of informatics, they are particularly suitable for solving problems that, in addition to being complex, are sometimes incompletely defined, or there is a vague and even stochastic relationship between the quantities involved. Neural networks build a mathematical model from a sample of real data, called "training data", to predict or make decisions without being explicitly programmed to do so.

For factors determining the final grade of each student, we choose the grades from the assessment components x_1, x_2, x_3, x_4. With the final grade G added to them, which is set by the teacher, we formed input-output samples of the type:

$$\{(x_{1i}, x_{2i}, x_{3i}, x_{4i}); G_i\}_{i=1}^{n} \tag{1}$$

In the first stage of our study, we used them to train several different types of artificial neural networks in Orange application.

At this stage, the modeled neural networks have different transfer functions in the hidden layer, and all other network parameters are the same (10 neurons in the hidden layer; trained with a quasi-Newtonian method for working with a limited amount of memory and using 1000 iterations).

The following transfer functions were studied:

- Linear: $f(x) = x$;
- Logistic: $f(x) = \frac{L}{1+e^{-k(x-x_0)}}$;
- Hyperbolic tangent: $f(x) = \frac{e^x - e^{-x}}{e^x + e^{-x}}$, and
- Rectified linear function: $f(x) = max(0, x)$

The artificial networks created at this stage were tested and compared with the "Test and Score" widget in Orange application, using this comparison indicators:

- Mean Squared Error: $MSE = \frac{1}{n} \sum_{i=1}^{n} (y_t - \hat{y}_t)^2$
- Root square of the mean square error: $RMSE = \sqrt{\frac{1}{n} \sum_{i=1}^{n} (y_t - \hat{y}_t)^2}$
- Mean Absolute Error: $MAE = \frac{\sum_{t=1}^{n} |y_t - \hat{y}_t|}{n}$
- Coefficient of determination: $R^2 = \frac{\sum_{i=1}^{n} (y_t - \hat{y}_t)^2}{\sum_{i=1}^{n} (y_t - \bar{y}_t)^2}$

In the indicated formulas, y_t denotes the actual data (grades), $\hat{y}_t$ are the data calculated by the regression model, $\bar{y}_t$ is the arithmetic mean of the data (grades) from the sample provided, and n is the sample size we have (the number of input-output samples for training). Here $(y_t - \hat{y}_t)^2$ are the squares of the differences in the interpolation of the respective nodes, and $(y_t - \bar{y}_t)^2$ are the squares of the remainder when calculating the coefficient R^2.

Table. 2. Evaluation results when using different transfer functions

Model	MSE	RMSE	MAE	R2
NN Identity L-BFGS-B	0.470	0.686	0.528	0.691
NN ReLu	1.206	1.098	0.900	0.207
NN Tanh	1.472	1.213	0.918	0.032
NN Logistic	3.476	1.864	1.113	−1.287

When using the "Test and Score 1" widget we found, that the **Neural Network Identity** network interpolates the least-error connections across most of the criteria used (Table 2). This shows that the teachers who assessed the students' work used a close to linear logic. R2 can be a negative number, when the model created by the neural network differs significantly from the real model. This may be a signal that a constant term should be added to the model. In Orange widgets, the logistic transfer function is built into neural network modules, and we can't change its expression.

This led us to use artificial neural networks with a linear transfer function of neurons.

In the second phase of our study, we compared the performance of neural networks trained on different algorithms. The comparisons were performed with an additional "Test and Score 2" widget and covered the following training methods:

- Quasi-Newton method for working with limited memory - L-BFGS-B;
- Stochastic gradient descent – SGD;
- The "Adam" training algorithm, as an extension of the stochastic gradient descent [16].

Test and Score 2 has shown a clear advantage in L-BFGS-B (MSE $= 0.462$) and "Adam" (MSE $= 0.472$) over the usual gradient descent, and with a slight advantage of the quasi-Newton method.

Experimenting with changing the number of neurons in the hidden layer and learning epochs did not lead to significant changes, which allows us to choose the ***Neural Network Identity L-BFGS-B*** for training in order to form a final assessment of the students. After the training of the network, conducted with 110 input-output samples of the type (1), a ***Test Variable*** is submitted to it, containing the results of the obtained knowledge of the students on the tests $x_1 - x_3$. and the assessment of exercises x_4. The number of examples to perform this test is 26.

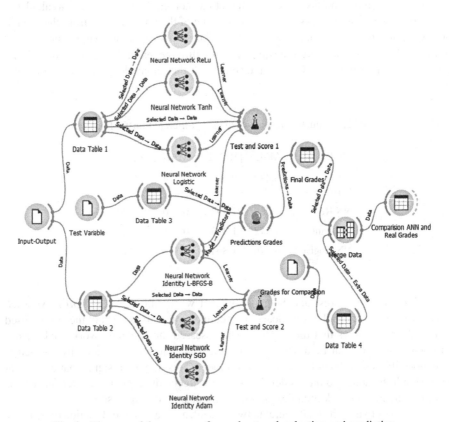

Fig. 1. Diagram of the process of neural network selection and predicting

The process of prediction from the ***Neural Network Identity L-BFGS-B*** network is implemented in the "Predictions Grades" widget. A file with the actual estimates loaded in the "Grades for Comparison" widget is used to estimate the prediction, and the differences between the predicted and the actual grades can be seen in "Comparison ANN and Real Grades" widget. A model of the experiment, neural network selection, prediction and error estimation is shown in Fig. 1. Taking into account the actual grades in all cases included in the test variable, it turns out that the MSE in assessing students' knowledge by the neural network has a value of 0.486.

3.3   Application of the SVM Method

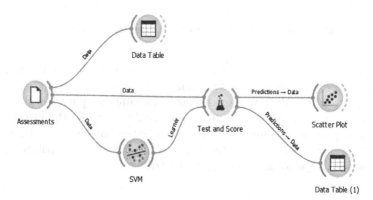

Fig. 2. Data analysis workflow using Support Vector Machines method

Support Vector Machine (SVM) is a supervised learning algorithm for machine learning. It is mainly used for classification problems but is also suitable for regression problems. It is supported in Orange application through the SVM widget. It allows predictions to be made, the accuracy of which depends on the specific settings.

The following widgets were used sequentially for the data workflow - File, SVM, Test and Score, and Scatter Plot (Fig. 2). The SVM widget supports two types of SVM - SVM and vSVM. The first one is applicable for classification and regression tasks, and the second - only for regression tasks. We have chosen the first type. Another important parameter is the Kernel, as Orange currently supports four options - Linear, Polynomial, RBF and Sigmoid. A Kernel is a function that transforms attribute space to a new feature space to fit the maximum-margin hyperplane.

Our experiments showed the best results with Linear Kernel. At set parameters - Cost = 0.10, Regression loss epsilon = 0.25, Numerical tolerance = 0.50 and Iteration limit = 100, we received the following errors: MSE = 0.115, RMSE = 0.039, MAE = 0.271 and R2 = 0.924.

3.4   Using the Linear Regression Method

Linear Regression is a machine learning approach that detects a linear dependence of a variable on one or more independent variables using already known data. Based on this dependence, given specific values of the independent variables, the algorithm predicts values for the target variable.

The data analysis workflow created by us is presented in Fig. 3. It includes widgets that perform the following activities sequentially - loading data, selecting training

Fig. 3. Data analysis workflow using Linear Regression method

data and validation data, defining features and target values, applying linear regression, predictions and visualizing the results as values of coefficients and charts.

When creating the model, we experimented with different options - with and without regularization. Some of the results are presented in Table 3. The model without regularization gives the best forecasts. The reported errors are MSE = 0.141, RMSE = 0.375, MAE = 0.312, R2 = 0.922. The results of this model are comparable to Ridge Regression, with Regularization strength from 0.0001 to 2.

Table 3. Main results when using the Linear Regression method

Learning model	Regularization strength	MSE	RMSE	MAE	R2
Ridge regression	0.8	0.141	0.375	0.312	0.922
Ridge regression	2	0.141	0.376	0.313	0.922
Lasso regression	0.03	0.144	0.379	0.316	0.920
Lasso regression	0.3	0.241	0.491	0.393	0.867
Elastic net regression	0.01	0.414	0.376	0.313	0.922
Elastic net regression	0.07	0.270	0.519	0.419	0.851

As expected, the results of the Linear Regression are very close to those of the SVM method. There is a natural mathematical explanation for this. Linear Regression makes it possible to analyze the influence of one or more independent variables on a dependent variable. The function expressing this dependence is of the type:

$$Y = a_0 + a_1x_1 + a_2x_2 + \cdots + a_nx_n + \varepsilon, \text{ where:}$$

a_i are regression coefficients and x_i are independent variables, i = 1, 2, ... n, and ε is an accidental error. The SVM method works in a similar way in the special case of working with close to linear logic.

4   Improving Results Using the GARP Method

The errors in our experiments with neural networks were not small enough. There are several reasons for this:

- Orange's standard NN tools have limited settings and control options.
- The small dataset of input-output samples does not allow good enough training of neural networks.
- The logic used by the assessor when forming the final assessments, in some cases uses various subjective factors related to the performance and achievements of the learner. Using floor or ceiling functions instead of rounding causes instability in the simple linear evaluation logic. The presence of "evaluation-exceptions" complicates the "unravelling" of the evaluator's close to linear logic by the artificial intelligence algorithms and their automatic transformation into a linear evaluation formula.

To solve the problem with the small number of input-output samples, we used the GARP method created by us [17]. In it, on the basis of a randomly selected part of the initial input-output samples, additional samples are generated, which can be used in the training of artificial intelligence algorithms. A condition for the successful use of the GARP method is the assumption that the estimation function is close to the linear one.

Experiments conducted with a dataset generated by the GARP method led to significantly better trained neural networks. For example, the error values for Neural Network Logistic are MSE = 0.011, RMSE = 0.106, MAE = 0.054 and R2 = 0.975. In addition to reducing the errors of MSE, RMSE and MAE, the application of GARP further brings the value of the coefficient of determination R2 closer to 1, emphasizing the linear logic captured by the neural network in calculating assessments. A comparative analysis of the experimented models is presented in Table 4.

Table 4. Comparison of the developed models

Model	MSE	RMSE	MAE	R2
Neural Network Identity L-BFGS-B	0.462	0.679	0.520	0.696
SVM Model	0.115	0.039	0.271	0.924
Linear Regression Model	0.141	0.375	0.312	0.922
Neural Network with GARP data	0.011	0.106	0.054	0.975

Improvements to the GARP method, as well as the development of tools for its integration into Orange, are the subject of further research.

5 Conclusion

The development and training of higher-order thinking skills should be a priority for every teacher. Assessing these skills requires significant preparation for developing and improving a system of various forms, methods and tools for testing and assessment. In some cases, students' achievements are difficult to assess with formulas, and teachers make a subjective assessment based on the relationships between the different assessment components assessing lower and higher-order thinking skills.

The article describes experiments for automated assessment of students' knowledge and skills using three methods of artificial intelligence - Neural Networks, SVM and Linear Regression. They use machine learning in which they build a mathematical model based on a sample of real data. They then calculate predictions for specific values of the independent variables.

The experiments aimed to show a way in which teachers can use different methods of artificial intelligence. The Orange application provides convenient machine learning tools that can be used by people without in-depth knowledge of programming and math. The three methods used in an appropriate way can form the final assessments of the students with a small error. Moreover, the predicted results generated by the algorithms can be used in another aspect – to detect incorrectly formed grades by the teacher. And yet, the final decision is up to the teacher.

Acknowledgements. The work is funded by the MU21-FMI-004 project at the Research Fund of the University of Plovdiv "Paisii Hilendarski".

References

1. Ruskov, S., Ruskova, Y.: The problem of evaluation in higher education. In: Proceedings of the XXI International Scientific Technical Conference Trans & MOTAUTO 2013, vol. 3, pp. 73–97 (2013). ISSN 1310-3946
2. Orange. https://orangedatamining.com/. Accessed 15 June 2021
3. Bloom, B., Engelhart, M., Furst, E., et. al.: Taxonomy of Educational Objectives: The Classification of Educational Goals. Handbook I: Cognitive Domain. David McKay Company, New York (1956)
4. Retnawati, H., Djidu, H., Kartianom, E., Apino, R.: Teachers' knowledge about higher-order thinking skills and its learning strategy. Prob. Educ. 21st Century **76**(2), 215 (2018)
5. Brookhar, S.: How to assess higher-order thinking skills in your classroom. ASCD (2010). ISBN: 978-1-4166-1048-9
6. Abosalem, Y.: Assessment techniques and students' higher-order thinking skills. Int. J. Second. Educ. **4**(1), 1–11 (2016)
7. Hadzhikolev, E., Hadzhikoleva, S., Yotov, K., Orozova, D.: Models for multicomponent fuzzy evaluation, with a focus on the assessment of higher-order thinking skills. TEM J. **9**(4), 1656 (2020)
8. Hadzhikolev, E., Yotov, K., Trankov, M., Hadzhikoleva, S.: Use of neural networks in assessing knowledge and skills of university students. In: Proceedings of ICERI 2019 Conference, 11–13 November 2019, Seville, Spain, pp. 7474–7484 (2019)
9. Kiryakova, G.: E-assessment – beyond the traditional assessment in digital environment. IOP Conf. Ser. Mater. Sci. Eng. **1031**,(2021). https://doi.org/10.1088/1757-899X/1031/1/01206
10. Johansson, E.: The assessment of higher-order thinking skills in online EFL courses: a quantitative content analysis. Nordic J. Engl. Stud. 224–256 (2020). https://doi.org/10.35360/njes.519
11. Serevina, V., Sariand, Y., Maynastiti, D.: Developing high order thinking skills (HOTS) assessment instrument for fluid static at senior high school. IOP Conf. Ser. J. Phys. Conf. Ser. **1185**,(2019). https://doi.org/10.1088/1742-6596/1185/1/012034
12. Stoitsov, G.: Assessment of the results from conducted experimental training in computer networks and communications in the laboratory exercises. TEM J. **6**(2), 185–191 (2017)

13. Hristov, H., Chochev, N.: Qualitative research of conference online learning for web design students. Math. Inf. **64**(2), 207–221 (2021)
14. Burkov, A.: The Hundred-Page Machine Learning Book (2019). ISBN: 10:199957950X
15. Alpaydin, E.: Introduction to Machine Learning. The MIT Press, Cambridge (2020). ISBN: 10:0262043793
16. Kingma, D., Ba, J.: Adam: a method for stochastic optimization. In: The 3rd International Conference for Learning Representations, San Diego (2015)
17. Yotov, K., Hadzhikolev, E., Hadzhikoleva, S.: GARP method – an approach to increasing the dataset for the training of artificial neural networks. Int. J. Adv. Trends Comput. Sci. Eng. **9**(5), 8972–8977 (2020). ISSN: 2278-3091

Predicting and Classifying Drug Interactions

Elena Stefanovska$^{(\boxtimes)}$ and Sonja Gievska$^{(\boxtimes)}$

Faculty of Computer Science and Engineering, Ss. Cyril and Methodius University,
Rugjer Boshkovikj 16, Skopje, Republic of North Macedonia
`elena.stefanovska@students.finki.ukim.mk`, `sonja.gievska@finki.ukim.mk`

Abstract. Experimentally identifying previously unknown drug-drug interactions (DDIs) that might cause potentially adverse drug reactions or alter drug's effectiveness when a combination of two or more drugs are used is a costly task. By contrast, many computational-algorithmic approaches have been used as a faster solution to the problem. Our current research efforts have been directed toward comparative performance evaluation of several approaches for discovering and classifying previously unknown interactions between two drugs. In this research, the task of discovering new DDIs have been formalized as a link prediction task in an interaction network constructed on the basis of previously known drug interactions. Several approaches for link prediction have been experimented with to find empirical evidence of their performance standing. Classifying the interaction type of a newly discovered DDI on the basis of the molecular compound of the drugs involved have also been explored.

Keywords: Drug-drug interaction · Link prediction · Node embeddings · Molecular fingerprints · Graph auto-encoder · Random walk with restart · GraphSAGE

1 Introduction

One of the critical requirements for advancing the field of drug discovery and development is identifying potentially harmful interactions between two or more drugs when they are co-administered to a patient. The challenge is to discover a drug-drug interaction (DDI) before the drug is approved to be used for therapeutic cure of a certain disease or a symptom [11]. Computational predictive methods are seen as a faster and safer alternative to the time-consuming and laborious process of identifying interactions between drugs. However, the performance of the methods experimented in the past is typically strongly affected by the quality of manually-engineered features. Various features that require domain-expert knowledge have been experimented with, from the molecular structure of a drug to its side effects.

The development of feasible computational methods to predict unknown association between two drugs have received a heightened attention in recent years.

© Springer Nature Switzerland AG 2022
L. Antovski and G. Armenski (Eds.): ICT Innovations 2021, CCIS 1521, pp. 26–37, 2022.
https://doi.org/10.1007/978-3-031-04206-5_3

While early DDI prediction methods have relied upon detailed information on the structural properties of the drugs and their targets i.e. proteins, most applications and research studies today rely on extracting information from publicly available repositories i.e. drug-drug interaction networks. Graph representation learning approaches based on random walk and deep learning in DDIs networks have emerged as a new trend. The methods evaluated in this paper fall under this category. The high-dimensional dense matrix of graph data are usually mapped to a low-dimensional vector space using some of the techniques for graph embeddings.

In the last decade, graph-based approaches have achieved unprecedented improvements in performance across a broad spectrum of problems ranging from recommender systems to learning molecular fingerprints. The interest graph-based analysis has attracted in the research on drug discovery have motivated our research in the direction of extracting knowledge from drug-drug and drug-food interaction networks to facilitate the discovery of new interaction that might have unwanted effects on patients.

What early research work in the field has in common is that information from a variety of sources were exploited for discovering new interactions between two drugs. In particular, similarity metrics based on local and global network structures in drug-drug interaction networks [10,14,20,21], chemical molecular drug compounds [12], drug targets, drug's side effects and other features have been frequently explored [22].

While similarity-based approaches [12,14,20,21] are still in use as baseline methods, representation learning on graphs, that is finding low-dimensional embeddings of nodes that preserve the similarity of nodes and structural properties of entire graphs are hailed as state-of-the art approaches to link prediction. In general, graph embedding techniques belong to three categories: factorization methods, random walk based [10] and deep learning based [11] methods. Identifying the type of a new computationally discovered drug-drug interaction usually accompanies or complements the task of DDI prediction. Knowing the interaction type of a DDI has a potential to enhance our knowledge and understanding of drug interactions, especially those DDI that might cause adverse drug reactions [11–13].

The objective of this research was two-fold: 1) to investigate the potential of several methods for predicting hypothetical i.e. unknown drug interactions, and 2) to explore different classification methods for identifying the type a given drug interaction belongs to. We focus our study pertaining to the first objective by formulating the problem of discovering new drug interactions as a link prediction. Link prediction refers to the task of predicting likely but missing links in a graph; for the problem at hand we use the graph that encodes information about previously confirmed interactions between drugs. Regarding our second research objective i.e. determining the type of a newly discovered drug-drug interaction, chemical fingerprint of the pairs involved are utilized.

Creation of benchmarking repositories containing previously confirmed drug-drug interactions, such as DrugBank[1] gold standard DDI dataset established by

---

[1] https://go.drugbank.com/.

Wishart et Al [19] gold standard DDI dataset, SemMedDB[2], and BioSNAP[3] have played an important role in advancing the field. A subset of 1024 drugs and their 70,000 interactions generated by a previous computational framework known as DeepDDI [12] have been targeted in our extensive experimentation. The chemical structure in SMILES format was extracted from the DrugBank.

After a brief summary of related research presented in Sect. 2, we introduce the datasets and present the details of the adopted methodology in Sect. 3. The findings and the interpretation of the results are presented in Sect. 4, while the last section concludes the paper and points to direction for future research.

2   Related Work

A review of current research on the topic of computational discovery of new DDIs have revealed common themes and differences, however the studies closely related to ours are discussed.

Drug-drug prediction has been previously approached as a link prediction problem using ensemble-based classifier and two novel matrix factorization methods as proposed by Shtar et al. in [14]. In the study, the authors have used the previous releases of the DrugBank dataset for training their predictive model that was subsequently tested on newer DataBank releases containing more drugs and interactions between drug pairs. Several similarity measures, such as Adamic-Adar, Jaccardi and Katz have been tested.

Several graph embeddings, including DeepWalk, node2vec, LINE and meta-path2vec were trained on a drug-target interaction network and later evaluated on several benchmark datasets for DDI prediction [11]. The findings of the study showed that the graph auto-encoder outperformed other classifiers for learning graph embeddings.

A computational framework DeepDDI that uses deep neural network for predicting drug reactions based on the names and structural properties of a pair of drugs is proposed in [12]. Previously confirmed drug reactions between drugs contained in the gold standard drug-drug interaction dataset, DrugBank [19] has been used as a basis for the prediction. Chemical structural information has been represented as a molecular fingerprint of the drug in a simplified SMILES format. The DeepDDI model predicts the type of a new DDI that could belong to one of the 86 interaction types with 92.4% mean accuracy. The output of the model is a human-readable sentence that specifies the pharmacological effects and/or adverse drug reactions between two drugs. The same predictive model DeepDDI has been used for predicting the type of interactions between a pair of drug and food constituent [12]. While similar in the objective to ours, our study differs in the methods used for prediction and classification of new drug-drug interactions and their type.

Many of the early studies on drug-drug and drug-target prediction task, suffered from the fundamental weakness of taking into account only the connectivity

---

[2] https://skr3.nlm.nih.gov/SemMedDB/.

[3] http://snap.stanford.edu/biodata/.

between pair of drugs without any regard for the global structural properties of the underlying network. Lee and Nam have proposed employing random walk with restart (RwR) to predict interactions between drugs and targets on the basis of the protein-protein interactome network [10]. The authors have pointed out that reweighting features by the RwR method has led to performance advantage compared to previous research methodologies.

Inspired by previous successes of using GraphSAGE for inductive learning of node representation [7] on similar tasks [5,6], we have also experimented with its applicability on our dataset comprised of DDIs augmented with molecular structural information on drug pairs.

DECAGON was proposed in [22] as a novel graph convolutional network for predicting polypharmacy side effects that can be used on large multimodal graphs that include various types of information on drugs, proteins and side effects, which are relevant for the task at hand. We are currently extending the information on which the new DDI prediction will rely, by incorporating and fusing information from a variety of sources i.e. existing repositories and knowledge graphs.

The approaches tested in this research have drawn upon the experiences of previous research on related problems and similar datasets.

3   Datasets

The first dataset was based on the network generated by a previous computational framework known as DeepDDI [12]. The output of DeepDDI is a human-readable sentence that describes the drug interaction between a pair of drugs if exists plus the number indicating the interaction type. Previously confirmed drug reactions between drugs contained in the gold standard drug-drug interaction dataset. DrugBank [19] has been used as a basis for the prediction. For our first objective of predicting new interactions between drugs, we have created a dataset that contains 70,000 DDIs generated by the DeepDDI model.

For each drug, its molecular structure has been converted into a molecular fingerprint that represents a one hot encoding vector. The binary value for each element represents the presence or absence of a particular substructure in a molecule. Graph Only Fingerprint (GraphFP), a specialized version of molecular fingerprint that uses 1024 bits to encode the structure of a drug without chemical bond information, has been used for generating our second dataset. The PaDEL-descriptor software was used to create the molecular fingerprints. This software uses the open source Chemistry Development Kit (CDK) [18] to represent the chemical concepts into suitable data structures that can be further manipulated. For each DDI interaction pair, the molecular fingerprint encodings of both drugs in a pair have been concatenated. By doing so, every interaction between a pair of drugs was represented by a vector of 2048 features. In addition, the interaction type was included in the molecular fingerprint feature vector. Graph fingerprints were retrieved in a simplified molecular input line entry system (SMILES) format from the DrugBank site, while the missing molecular structures were manually extracted from the PubChem site[4]. SMILES format represents the chemical structure in a machine-readable form as a concatenated list of symbols for atomic elements and symbols for the type of bond between them.

4 Methodology

The task of predicting new and unknown drug-drug interactions has been defined as a link prediction on the graph representing the information on previously identified drug interactions. The nodes in the graph represent drugs and the edges between them denote their known interactions. An edge between two nodes was created based on the information in our first dataset containing 1123 drugs and roughly 70,000 interactions between drug pairs. The dataset was split into a training and a testing dataset; 10% of the edges representing existing interactions have been removed to serve as positive samples. An equal number of negative samples were randomly chosen from the set of non-existent edges.

Several approaches have been explored, including similarity-based, random walk-based and deep graph convolutional link prediction methods.

4.1 Similarity Based Link Prediction Methods

The basic formulation of the similarity-based link prediction task between a pair of nodes u and v, $u, v \epsilon$ V, in a graph G, is given by a score $S(u, v)$. The underlying assumption according to [2], is that a pair of nodes with higher similarity scores have more chances to be connected i.e., an edge between them exists, compared to the ones with lower values. The existing nodes and edges in a graph, in our case the drug-drug interaction network, are used to calculate the similarity scores of the nodes and predict the possibility of existence of new links that are more likely to happen than others.

We have evaluated three similarity-based methods, namely Jaccard coefficient, Adamic-Adar index and Preferential attachment score that have been widely used in previous research on a number of tasks [2]. Similarity metrics, Jaccard and Adamic-Adar use the similarity of the local neighborhood of a pair of nodes to infer the probability of them being connected by an edge. In contrast, the metrics such as the Preferential attachment score use the global structural properties of the network by assigning similarity scores to each node.

Jaccard coefficient between two nodes u and v is calculated as the ratio of the number of the neighboring nodes they share compared to the total number of their neighboring nodes. Given two nodes u, v ϵ V, the Jaccard coefficient is calculated by the following formula:

$$S^{Jaccard}(u, v) = \frac{|\Gamma(v) \cap \Gamma(u)|}{\Gamma(v) \cup \Gamma(u)} \tag{1}$$

where $\Gamma|u|$ and $\Gamma|v|$ represent the sets of neighbours of nodes u and v, respectively.

---

[4] https://pubchem.ncbi.nlm.nih.gov/.

Adamic-Adar (AA) index is a topology-based similarity measure, initially proposed to predict connections in social networks based on the degree centrality of the common neighbors two nodes share. It is calculated as the sum of the inverse degree centrality of two nodes, with an implication that the more friends two nodes have in common, the lower the AA index value will be. The formula for calculating the AA index between two nodes u and v follows [1]:

$$S^{AA}(u,v) = \sum_{w \in \Gamma(v) \cap \Gamma(u)} \frac{1}{log|\Gamma(w)|} \tag{2}$$

where, $\Gamma|w|$ is the set of adjacent nodes of w, which is a common neighbour of the node u and node v.

The underlying premise of the preferential attachment score algorithm is that the more connected a node is, the more likely the node will establish new links to other nodes [3]. The similarity score, based on the preferential attachment algorithm takes into account the degree of both nodes in the pair under investigation, as shown in formula 3. The computational complexity of calculating the scores on a global scale for a given network is the major drawback of this similarity-based measure.

$$|\Gamma(u)| * |\Gamma(v)| \tag{3}$$

4.2   Random Walk-Based Link Prediction

Network-based methods for link prediction, such as random walks are based on establishing a suitable ranking algorithm for graph entities that is used for prediction of new links. Random walk with restart (RwR) is one of the most popular network propagation algorithm that has shown to be particularly applicable to biomedical problems involving graph-based knowledge [2,15].

Given a connected weighted graph G(V, E) with a set of nodes V = $v_1, v_2, ...,$ v_N and a set of links E = $\{(v_i, v_j) - v_i, v_j \in$ V$\}$, a set of source/seed nodes S $\subseteq$ V and a $N \ x \ N$ adjacency matrix W of link weights are defined. Random walk with restart is a variation of the original random walk algorithm [16] that assumes a random walker moves from a current node to a randomly selected adjacent node or going back to a seed node source nodes with a back-probability $\gamma(0, 1)$. The RwR algorithm can be formally defined using the following notations [9]:

$$p^{t+1} = (1 - \gamma)W^t p^t + \gamma p^0 \tag{4}$$

p^t is a $N \ x \ 1$ vector of probabilities for each of the $|V|$ nodes at a time t, where the i-th element denotes the probability that the walker is currently at node v_i $\in$ V, and p^0 is the $N \ x \ 1$ initial vector of probabilities defined as follows [9]:

$$p^0 = \begin{cases} \frac{1}{|S|}, & \text{if } v_i \in S. \\ 0, & \text{otherwise.} \end{cases} \tag{5}$$

A graph transition matrix W' is defined as follows [9]:

$$(W')_{ij} = \frac{(W)_{ij}}{\sum_j (W)_{ij}} \tag{6}$$

where each element in W' denotes a probability that a walker at v_i moves to v_j among $V \setminus \{v_i\}$.

4.3   Deep Learning Link Prediction Methods

Motivated by previous success of deep neural networks, in particular Graph Auto-Encoder presented in [8], for learning node and graph representation, we have also evaluated two deep learning approaches on the task of predicting new drug interactions.

Graph Auto-Encoder consists of graph convolutional encoder and decoder networks, that learn in an unsupervised manner the low-dimensional representation of node embeddings. The output of the decoder reconstructs the adjacency matrix of a given graph, G.

Let $a_i \in R^N$ be an adjacency vector i.e., a row in an adjacency matrix A of a graph G, which represents the local neighborhood of the i-th node. A four-layer auto-encoder architecture we have used to learn the node embeddings from local neighborhood structures is known under the name of LoNGAE (Local Neighborhood Graph Auto-encoder) [17]. The LoNGAE neural model includes a set of non-linear transformations on a_i by its two components: an encoder g(a_i): $R^N \rightarrow R^D$, and decoder f(z_i): $R^D \rightarrow R^N$. A two-stacked-layers encoder derives a D-dimensional latent feature representation of the i-th node $z_i \in R^D$, while the two-stacked-layers decoder outputs an approximate reconstruction of the vector, $\hat{a}_i \in R^N$. The reconstructed output $\hat{a}$ represents the likelihood score for existence of an edge between two nodes. The hidden distributions for the encoder and decoder are given with the following [17]:

$$z_i = g(a_i) = ReLU(W \cdot ReLU(Va_i + b^{(1)}) + b^{(2)}) - EncoderPart, \tag{7}$$

$$\hat{a}_i = f(z_i) = V^T \cdot ReLU(W^T z_i + b^{(3)}) + b^{(4)} - DecoderPart, \tag{8}$$

$$\hat{a}_i = h(a_i) = f(g(a_i)) - Auto-encoder. \tag{9}$$

The activation function used in our model implementation was the rectified linear unit, ReLU(x) = max(0, x) . The auto-encoder is constrained to be symmetrical with shared parameters for W, V between the encoder and decoder, resulting in almost two times fewer parameters compared to an unconstrained architecture. Note that the bias units denoted by b do not share parameters. $\{W^T, V^T\}$ are transposed versions of $\{W, V\}$. The parameters to be learned are summarized as $\theta = \{W, V, b^{(k)}\}$, where k = 1, 2, 3, 4.

A graph-convolutional network (GCN) has been used for learning node representations, for which the initial graph of DDIs was augmented with additional node and edge features i.e. molecular fingerprint of each drug pair and the interaction type accordingly. A variation of Graph Convolutional Networks, very similar to the implementation of GraphSAGE framework [7] was used for the task of predicting drug-drug interactions in [6]. GraphSAGE is framework that has proved to be effective for training on large scale networks to inductively learn the embeddings by its sample and aggregate approach. Sampling refers to the process of drawing node's neighborhoods of a certain depth k, while an set of aggregator functions incrementally, with each iteration, aggregate features from node's neighborhoods. The aggregator functions, denoted as $AGGREGATE_k$, $\forall k \,\epsilon\, \{1, \ldots, K\}$ are jointly learned with the weight matrices W_k, $\forall k \,\epsilon\, \{1, \ldots, K\}$ by propagating information to node's neighbors at various depth. The embeddings generation process is described by the following equations [7]:

$$h^k_{N(v)} \leftarrow AGGREGATE_k(\{h^{k-1}_u, \forall u \in N(v)\}) \tag{10}$$

$$h^k_v \leftarrow \sigma\left(W^k \cdot CONCAT(h^{k-1}_v, h^k_{N(v)})\right) \tag{11}$$

where h^k is the node representation at depth k, $v, u \in V$, $N : v \rightarrow 2^n$ is a neighbourhood function, σ denotes the non-linearity.

Standard stochastic gradient descent and backpropagation techniques are used for learning in a fully unsupervised manner. The graph based loss function is given by the following formula [7]:

$$J_G(z_u) = -log(\sigma(z^T_u z_v)) - Q \cdot E_{v_n \sim P_n(v)} log(\sigma(z^T_u z_{v_n})), \tag{12}$$

where v is a node that appears near node u on fixed-length random walk, σ is the sigmoid function, P_n is a negative sampling distribution, and Q defines the number of negative samples.

There is no ordering of nodes in a graph, so the aggregation function needs to be symmetric. Three aggregator functions have been examined in the original proposal: mean aggregator, LSTM-based (Long Short Term Memory) and pooling aggregator function, with pooling aggregator showing the best results [7].

Our GraphSAGE model was consisted of two GraphSAGE layers, and was implemented using Mean aggregator, as set by default. The embeddings of the nodes within a drug pair were combined by inner product function in order to obtain the final prediction whether a link between them is predicted to exist or not.

4.4   Drug-Drug Interaction Type Classification

The second task we have posed in this research was to identify the interaction type of a newly discovered drug interaction, for which the concatenated feature vectors consisted of molecular structural information have been used. Three methods we have chosen to tackle the second problem of determining the interaction type of a new DDI belong to both, traditional machine learning and deep learning approaches. The traditional classifiers we have chosen as suitable for the task were: k-nearest neighbors (KNN) classifier and an ensemble classifier. The rationale for the first relates to utilizing the similarity of feature vectors as a distance measure, while XGBoost [4] was used as an ensemble classifier because of their superiority in performance across a variety of tasks [4]. The longer training time of the ensemble classifier was expected. A simple deep LSTM (Long Short Term Memory) neural network, implemented as a 6–8 layers architecture including dense and dropout layers with around 300,000 parameters was also employed for classifying the drug-drug interaction types. The second dataset we have used that incorporates a vector of 2049 features as a molecular fingerprint for each pair of drugs is rather unbalanced; the number of positive/negative samples varies across all 86 interaction types (classes).

5   Discussion of Results

5.1   Prediction of Drug-Drug Interactions

Table 1 shows the performance metrics of the similarity-based link prediction methods, namely the accuracy, ROC, precision, recall and F1 that are macro-averaged. The performance analysis of link prediction methods used for discovering new drug-drug interactions have pointed out the advantages and limitations of different models we have employed on the task. The hyperparameters were experimentally set, i.e. the threshold value for the Jaccard coefficient, Adamic-Adar Index and Preferential attachment score was set to 0.001, 0.05 and 150, respectively. Preferential attachment score has obtained the best performance results, with an overall accuracy of 0.69, although it should be noted that the time complexity of this method depends on the size of the network and might be of a great concern for large scale networks. The performance results vary across classes and we could speculate that the unbalanced dataset used in the experimental evaluation has a degrading effect on the performance.

Table 1. Performance results of similarity-based link prediction methods on the task of predicting new drug interactions.

Performance measures					
Methods	Accuracy	ROC	Precision	Recall	F1-Score
Jaccard Coefficient	0.51	0.51	0.52	0.51	0.45
Adamic-Adar Index	0.51	0.51	0.52	0.51	0.45
Preferential Attachment score	0.69	0.69	0.78	0.69	0.67

Table 2 presents the comparative performance results of the four different types of methods, Preferential attachment score as the best performing similarity-based link prediction method, random walk-based, Graph Auto-Encoder and GraphSAGE method. The performance results strongly demonstrate that the random ralk with restart model has outperformed the other methods achieving the best ROC value of 0.9 and an average precision score of 0.84. The obtained value for the macro-averaged recall was 0.90, which points out that the RwR model handles better the variance in data. Little evidence was found for the advantages of the graph auto-encoder model compared to the RwR model. We might speculate that augmenting the models like RwR and Graph Auto-Encoder with additional features for drug pairs and not relying only on the network topological properties could lead to some performance gains. Our implementation of GraphSAGE has achieved accuracy value of 0.76, making it the second best model, after RwR in our case. A deeper neural structure of the GraphSAGE model which would require more powerful processing resources is expected to achieve significantly higher performance values. It should be noted that even though this deep learning model was modest in size with around 16 000 parameters, it has achieved very encouraging results.

Table 2. Comparative performance results of the best performing similarity-based method, random walk with restart (RwR), graph auto-encoder (Graph AE) and Graph-SAGE on the task of predicting new drug interactions.

Performance measures					
Methods	Accuracy	ROC	Precision	Recall	F1-Score
Preferential Attachment score	0.69	0.69	0.78	0.69	0.67
Random Walks (RwR)	0.90	0.90	0.90	0.90	0.90
Graph AE	0.62	0.62	0.79	0.62	0.56
GraphSAGE	0.76	0.76	0.78	0.76	0.76

5.2   Classifying the Type of Drug-Drug Interaction

The second equally important objective in our study was to evaluate several methods for classifying a newly discovered drug-drug interaction type. Surprisingly, the simple LSTM deep learning network did not lead to any performance gains compared to other methods. Further improvements might be expected for batch sizes larger than 100. Our current research efforts are focused on hyperparameter optimization of the LSTM neural network.

The performance of the KNN model were not satisfactory, slightly above 0.54 for k = 3 neighbors. According to the results summarized in Table 3, the XGBoost ensemble model has obtained an overall accuracy of 0.63. Considering that the dataset used for interaction type classification did not have a uniform class distribution, the varying performance values, especially the precision and recall, were expected. We are currently attempting on improving the quantity and the quality of the data samples to improve the performance on this task.

Table 3. Performance results of the k-nearest neighbor and XGBoost on the task of interaction type classification.

Performance measures				
Methods	Accuracy	Precision	Recall	F1-Score
k-Nearest Neighbours	0.54	0.26	0.16	0.18
XGBoost	0.63	0.63	0.63	0.58

6   Conclusions

Undesirable effects of a medication that occur when simultaneously two or more drugs are administered to a patient need to be identified promptly. Improving our knowledge of potentially dangerous drug interactions is a recurring challenge in the field of drug discovery and development.

Due to considerations of scale, experimental design, and the cost of experimentally verifying interactions between drugs, it is safe to posit that most unknown interactions are unlikely to be experimentally characterized; instead computational and machine learning methods have been successfully deployed. Following the results of previous research in the field, we have experimented with and evaluated several models for the task of identifying new drug interactions and classifying their interaction type. Random walk with restart algorithm, performed best on the task of discovering drug-drug interactions, while the XGBoost ensemble model performed best on the task of classifying the drug-drug interaction type. The GraphSAGE model has achieved promising results on the task of drug interaction prediction and should be considered a method of choice when training of large scale dataset that require significant processing power are available.

Our current research efforts are directed towards extending our models and feature representation learning by including other relevant drug-related attributes, such as drug descriptions, currently available information on potential side effects, etc.

Acknowledgement. This work was partially financed by the Faculty of Computer Science and Engineering at the "Ss. Cyril and Methodius" University.

References

1. Adamic, L.A., Adar, E.: Friends and neighbors on the web. Soc. Netw. **25**(3), 211–230 (2003)
2. Backstrom, L., Leskovec, J.: Supervised random walks: predicting and recommending links in social networks. In: Proceedings of the fourth ACM International Conference on Web Search and Data Mining, pp. 635–644 (2011)
3. Barabási, A.L., Albert, R.: Emergence of scaling in random networks. Science **286**(5439), 509–512 (1999)
4. Chen, T., He, T., Benesty, M., Khotilovich, V., Tang, Y., Cho, H., et al.: XGBoost: extreme gradient boosting. R Package Version 0.4-2 **1**(4), 1–4 (2015)

5. Cui, C., et al.: Drug repurposing against breast cancer by integrating drug-exposure expression profiles and drug-drug links based on graph neural network. Bioinformatics **37**(18), 2930–2937 (2021)
6. Feeney, A., et al.: Relation matters in sampling: a scalable multi-relational graph neural network for drug-drug interaction prediction. arXiv preprint arXiv:2105.13975 (2021)
7. Hamilton, W.L., Ying, R., Leskovec, J.: Inductive representation learning on large graphs. In: Proceedings of the 31st International Conference on Neural Information Processing Systems, pp. 1025–1035 (2017)
8. Kipf, T.N., Welling, M.: Variational graph auto-encoders. arXiv preprint arXiv:1611.07308 (2016)
9. Le, D.H.: Random walk with restart: a powerful network propagation algorithm in bioinformatics field. In: 2017 4th NAFOSTED Conference on Information and Computer Science, pp. 242–247. IEEE (2017)
10. Lee, I., Nam, H.: Identification of drug-target interaction by a random walk with restart method on an interactome network. BMC Bioinform. **19**(8), 9–18 (2018)
11. Purkayastha, S., Mondal, I., Sarkar, S., Goyal, P., Pillai, J.K.: Drug-drug interactions prediction based on drug embedding and graph auto-encoder. In: 2019 IEEE 19th International Conference on Bioinformatics and Bioengineering (BIBE), pp. 547–552. IEEE (2019)
12. Ryu, J.Y., Kim, H.U., Lee, S.Y.: Deep learning improves prediction of drug-drug and drug-food interactions. Proc. Natl. Acad. Sci. **115**(18), E4304–E4311 (2018)
13. Seo, M., Shin, H.K., Myung, Y., Hwang, S., No, K.T.: Development of natural compound molecular fingerprint (NC-MFP) with the dictionary of natural products (DNP) for natural product-based drug development. J. Cheminform. **12**(1), 1–17 (2020)
14. Shtar, G., Rokach, L., Shapira, B.: Detecting drug-drug interactions using artificial neural networks and classic graph similarity measures. PLoS ONE **14**(8), e0219796 (2019)
15. Tong, H., Faloutsos, C., Pan, J.Y.: Fast random walk with restart and its applications. In: Sixth International Conference on Data Mining (ICDM 2006), pp. 613–622. IEEE (2006)
16. Tong, H., Faloutsos, C., Pan, J.Y.: Random walk with restart: fast solutions and applications. Knowl. Inf. Syst. **14**(3), 327–346 (2008)
17. Tran, P.V.: Learning to make predictions on graphs with autoencoders. In: 2018 IEEE 5th International Conference on Data Science and Advanced Analytics (DSAA), pp. 237–245. IEEE (2018)
18. Willighagen, E.L., et al.: The chemistry development kit (CDK) v2. 0: atom typing, depiction, molecular formulas, and substructure searching. J. Cheminform. **9**(1), 1–19 (2017)
19. Wishart, D.S., et al.: DrugBank 5.0: a major update to the DrugBank database for 2018. Nucleic Acids Res. **46**(D1), D1074–D1082 (2018)
20. Zhang, W., Chen, Y., Liu, F., Luo, F., Tian, G., Li, X.: Predicting potential drug-drug interactions by integrating chemical, biological, phenotypic and network data. BMC Bioinform. **18**(1), 1–12 (2017)
21. Zhang, W., et al.: SFLLN: a sparse feature learning ensemble method with linear neighborhood regularization for predicting drug-drug interactions. Inf. Sci. **497**, 189–201 (2019)
22. Zitnik, M., Agrawal, M., Leskovec, J.: Modeling polypharmacy side effects with graph convolutional networks. Bioinformatics **34**(13), i457–i466 (2018)

SemanticStyleGAN: Generative Image Inpainting Using Style-Based Generator

Darko Filipovski[✉] and Sonja Gievska

Faculty of Computer Science and Engineering, University of Ss. Cyril and Methodius in Skopje, Ruger Boskovic 16, 1000 Skopje, North Macedonia
darko.filipovski@students.finki.ukim.mk, sonja.gievska@finki.ukim.mk

Abstract. Image inpainting, a task of reconstructing missing or corrupted image regions, has a great potential to advance the fields of image editing and computational photography. Despite being a difficult problem, the researchers have made headway into the field thanks to the progress in representation learning with very deep convolutional neural networks and the successes in generation of realistic images by using Generative Adversarial Networks (GANs). To avoid the problems of generating blurry and distorted regions as well adding artefacts, a new model for image inpainting is proposed in this paper. The GAN-based architecture incorporates two frameworks, the model for semantic image inpainting and the models for style transfer, which have been pointed out to be successful in previous research. Our model was evaluated on the Flickr-Faces-HQ dataset. The results are promising and point out that using a combination of various GAN-based technologies could improve the performance on the task of image inpainting. Directions for future research are also discussed.

Keywords: Image inpainting · Generative adversarial networks · Style transfer

1 Introduction

Sophistication of machine vision that matches human visual perception skills has yet to be realized. Evolutionary development of our visual perception have equipped us to reach beyond the information gathered from our senses. We are capable of suppling the missing parts ourselves when perceiving and interpreting a visual scene, relying on our prior experiences. However, the success of state-of-the-art pretrained deep learning networks and generative adversarial networks have opened new promising avenues for the future to be explored.

Computational image inpainting, formulated as a task of reconstructing image regions being missed or corrupted has an enormous potential to advance the fields of object recognition, image editing and computational photography. Early methods for image inpainting were based on image models and extraction of local features of the area surrounding the missing region [3,5]. Finding the

L. Antovski and G. Armenski (Eds.): ICT Innovations 2021, CCIS 1521, pp. 38–51, 2022.
https://doi.org/10.1007/978-3-031-04206-5_4

best match for the missing region from the original image [2,4] or from a collection of images [7] have also been explored with varying success depending on the position of the corrupted region and/or the presence of texture or a pattern.

While reconstructing small image regions might be considered a reachable goal, replacing larger regions with perceptually and semantically plausible ones proves to be a challenging task. One explanation for the lower performance of early methods on this task is that image inpainting is not about modeling the local context surrounding the corrupted patch. The reconstruction should be guided by the wider context and the compositional structure of the original image. Despite being a difficult problem, the research community has recently made headway into this area thanks to the progresses in representation learning with very deep convolutional neural models (CNNs) [13,18] and Generative Adversarial Networks (GANs). The latter ones are widely regarded for their great ability to generate realistic images conditioned on previously learned context [6,12,17].

Our research follows the line of previous work on image inpainting using deep generative models. In particular, we have used our own implementation of the architecture for semantic image inpainting proposed in [20] as a baseline model. The process of training can be viewed as finding an encoding of the corrupted image that is closest to the latent representation of the image. This is achieved by defining a weighted context loss, obtained from the corrupted image, which acts as a condition, and a prior loss that penalizes unrealistic results. We have strengthen the baseline model by incorporating a modified version of the generator and discriminator proposed in the StyleGAN model [12]. Our proposed SemanticStyleGAN architecture was evaluated on the Flickr-Faces-HQ (FFHQ)[1] dataset containing 70,000 high-resolution images of human faces. The obtained results are encouraging and the insights into the unsuccessful reconstructions point out to few directions for future improvement.

We first give a brief overview of the research and methods most related to our approach in Sect. 2, before introducing the SemanticStyleGAN architecture advanced in this research in Sect. 3. Section 4 gives a description of the dataset on which our model was trained and evaluated, the training setup and the results of the experiments. The concluding section summarizes the research presented in this paper and points to few avenues for future research.

2   Related Work

Current technological trends in deep learning has real potential for advancing the endeavor in image inpainting. In this brief overview, we discuss several studies that are closely related to the method proposed in this research.

The Context Encoder model for semantic image inpainting presented in [15] consists of an encoder that learns the context of the corrupted image and a decoder that produces the missing region based on the encoded context. The

---

[1] https://github.com/NVlabs/ffhq-dataset.

network is trained in an unsupervised manner, minimizing the joint loss that includes: a reconstruction (L2) loss to capture the missing region structure, and an adversarial loss to facilitate generation of more realistic images. One of the limitations of this work is that its performance is highly dependable on the shape of the masks being used during training. Using arbitrary masks (dropped out region) might result in producing artifacts. To mitigate the problems related to generation of unwanted artifacts or distorted and blurry patches that are not consistent with the surrounding regions, a feed-forward generative network with novel contextual attention layer have been proposed in [21]. The contextual attention layer enables the system to attend to and copy the surrounding patches that are the best match for the features of the missing region. Attention propagation is implemented to maintain consistency of the generated missing region with the surroundings. The integrated network uses two encoders trained in parallel [21]. The first encoder is a simple dilated convolutional network that produces a rough estimate of the missing region, while the second encoder using contextual attentions attends to the most relevant surrounding patches using them as convolutional filters to further improve the initially generated patch. The aggregated output of the two encoders are fed into a decoder. The integrated neural model uses several techniques, such as spatially discounted reconstruction loss and two Wasserstein GAN losses [1] to improve the stability during training.

EdgeConnect [14] is an adversarial model that performs image inpainting in two stages. In the first pass the edges of the missing region are generated. In the subsequent stage, the edges are combined with finer details, such as color, texture of the rest of the image to complete the missing region generation. The generators are built based on the architectures presented in [11], that are commonly used for style transfer and super-resolution. As a discriminator, an architecture similar to PatchGAN [10] is employed. These new methods provide the means to capture complex information from images, thus producing superior performance results.

The deep generative model for semantic image inpainting proposed in [20] was used as a basis for the model proposed in this paper. The novelty of the approach for image inpainting advanced in [20] lies in using a trained Deep Convolutional Generative Adversarial Network (DCGAN) [17] to search for the closest encoding of the corrupted image in the latent image manifold. The proposed model is independent of the choice of the GAN variant architecture being used.

The popularity of GAN style transfer has attracted increasing attention, and many efforts have been taken to address the problem with a reciprocal impact on improving the generative adversarial architecture. One such variant, StyleGAN [12] incorporates a novel style-based generator architecture to improve the representation of high-level objects and better disentanglement of factors of variations in an intermediate latent space. We have utilized the strengths of StyleGAN as an extension to our baseline model inspired by the framework for semantic painting introduced in [20].

3   SemanticStyleGAN Architecture: Semantic Inpainting by Constrained Image Generation

In this paper, we introduce a novel GAN-based architecture, SemanticStyleGAN for image inpainting that is based on semantic image inpainting as suggested in [20] and incorporates a style-based GAN network to facilitate higher perceptual quality and easier separation of high-level objects when reconstructing a corrupted image. Our proposed approach to the problem of image inpainting utilizes a StyleGAN network [12] for an improved scale-specific control of the image synthesis process. By doing so, the inpainting process takes advantage of the generator G and the discriminator D trained on uncorrupted data.

Generative Adversarial Networks (GANs) introduced by Goodfellow et al. [6] is a powerful generative model that includes a pair of neural networks, a generator and a discriminator, simultaneously trained in an adversarial manner. Formally, the generator G is a differentiable function which maps an input variable z, sampled from a prior distribution p_z, to $G(z; \theta_g)$ from p_{model}. In contrast, the discriminator D is a simple classifier, which classifies samples as fake if they are generated from the generator's p_{model} distribution, and as real if they are sampled from the real training data p_{data}. More specifically, $D(x)$ is a scalar value, which represents the probability that x is sampled from p_{data} rather than p_{model}. These models are typically represented by Multilayer Perceptrons with parameters θ_g, for the generator, and θ_d for the discriminator.

When employed for the task of image inpainting, we have followed the method for semantic inpainting proposed in [20], which tries to find an image closest to the corrupted one by searching through the latent space. Learning an encoding $\hat{z}$ close enough to the original image is performed by "searching" the latent manifold while updating z.

The problem of finding $\hat{z}$ is formulated as an optimization problem i.e., minimizing an objective loss function with respect to z. This is formulated as

$$\hat{z} = \arg \min_{\mathbf{z}} \mathcal{L}_c(z|y, M) + \mathcal{L}_p(z), \qquad (1)$$

as suggested by [20], where $\mathcal{L}_c$ is the context loss of z given the corrupted image y and a mask M, whereas, $\mathcal{L}_p$ denotes the prior loss.

3.1   Context Loss

The context of the missing region is captured by a context loss, which is calculated by taking advantage of the remaining "uncorrupted" pixels in an image y. A suitable weighting scheme is employed, giving more weight to the pixels surrounding the corrupted region. Following [20], the importance weighting term for each pixel W_i is given by

$$W_i = \begin{cases} \sum_{j \in N(i)} \frac{(1-M_j)}{|N(i)|} & if \ M_i \neq 0 \\ 0 & if \ M_i = 0 \end{cases} \qquad (2)$$

where W_i denotes the importance weight of a pixel at location i, while $N(i)$ refers to the set of neighboring pixels of i in a local window of a particular size. During computation of the weighting term, the set of neighbors taken into account increases with the increase in the image resolution. An illustration of the importance weighting term, W, is shown in Fig. 1.

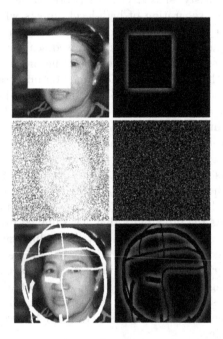

Fig. 1. Importance weighting matrix **W** derived from the corrupted image. (Left) Corrupted image. (Right) Importance weights. Grayscale values closer to 1 (white) indicate higher importance, while values closer to 0 (black) indicate lower importance.

The context loss is defined as a weighted l_1-norm difference between the generated image and the uncorrupted portion of the input image, as shown in Eq. 3 [20].

$$\mathcal{L}_c(z|y, M) = \|\mathbf{W} * (G(z) - y)\| \tag{3}$$

where * denotes element-wise multiplication.

In the case of DCGAN, the generated image $G(z)$ is obtained as a result of direct projection and upscaling of the input vector z. Consequently, changes to specific parts of z do not directly correspond to high-level features in the generated image. By contrast, introduction of the style-based generator [12] enables control of image synthesis via modifications to the style. More specifically, the effects of modifying specific styles are localized in the network and can affect only certain aspects of an image. In our SemanticStyleGAN model, the focused control over the importance of image features is performed by the StyleGAN network that we have employed as a replacement of DCGAN. The basic architecture of the StyleGAN is presented in Fig. 2.

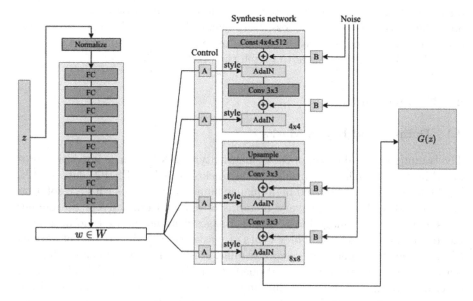

Fig. 2. StyleGAN architecture highlighting the control we affect by walking over the latent manifold

3.2   Prior Loss

Context loss takes into consideration individual pixel values with the intention to bring the resulting image pixels, which correspond to the uncorrupted pixels in y, closer to the intended value. However, the generated image may not look realistic at all, even though the pixel values are close to the original ones, and as a result an introduction of prior loss is needed. The prior loss $\mathcal{L}_p$ guarantees that the resulting image is generated from p_{data} distribution and as a result penalizes unrealistic images. $\mathcal{L}_p$ is taken to be the GAN's loss for training the discriminator D, which takes advantage of the discriminator's ability to differentiate between a generated and a real image, therefore the prior loss takes the form:

$$\mathcal{L}_p = \lambda \log\left(1 - D(G(z))\right) \tag{4}$$

as proposed by [20], where λ balances the two losses. In our implementation λ is set to 0.003.

3.3   Image Completion

The final step in obtaining the image with a reconstructed region that was previously missing or corrupted can be formulated as an optimization problem.

Initially, z is drawn from a uniform distribution and is passed through the generator G. The loss is applied to the output and the error is propagated backwards. These steps of generating outputs and backpropagating the error are repeated several times, in the process of obtaining the output $\hat{z}$. Generating $G(\hat{z})$ produces an output, which is expected to be close to the uncorrupted portion of an image y, controlled by the loss, from which the previously corrupted pixels can be extracted and replaced within y. Poisson blending as suggested in [16] is applied to the resulting image to match the pixel intensities.

4 Experiments

4.1 Dataset and Image Masks

The training and evaluation of the SemanticStyleGAN for image inpainting proposed in this research is performed on the Flickr-Faces-HQ (FFHQ) dataset [12]. The dataset consists of 70,000 high-quality human face images with a resolution of 1024×1024. The images vary in terms of age and ethnicity of the human faces as well as the image backgrounds. Images were normalized, no data augmentation have been used. Three types of masks were used to represent various types of realistic application usages: regular, noise and irregular masks. Regular masks are centered rectangles positioned at random locations within the image, covering 10% to 40% of the original image area. Noise masks are sampled from a "continuous uniform" distribution over half-open interval [0.0, 1.0) and cover 80% of the total image area. Irregular masks are extracted from the Quick Draw Irregular Mask dataset [9], which contains 50 million strokes drawn by human hand.

4.2 Implementation Details and Training Setup

The SemanticStyleGAN incorporates a style-based GAN, as a drop-in replacement for DCGAN, thus requiring a separate training process for the purpose of evaluation. The generator of the DCGAN is trained on 64×64 pixel resolution images with a batch size of 256. Optimization is performed using Adam optimizer, with a learning rate of $\alpha = 0.0002$ and coefficients $\beta_1 = 0.5$, $\beta_2 = 0.999$. We have adopted a pretrained version of the StyleGAN model from previous research [12], that was trained for 140,000 iterations over the FFHQ dataset, generating samples with a resolution of 256×256.

DCGAN based implementation samples a 100-dimensional vector from a standard normal distribution. A window size of 7 is used for generating the importance weighting term W. Obtaining the encoding $\hat{z}$ is performed by backpropagation of the loss for 1500 iterations. The SemanticStyleGAN component uses 512-dimensional vector sampled from a standard distribution. Context loss is computed using a weighting term, which captures a window of size 28. The loss is backpropagated for 3000 iterations. Differences in the window size for the weighting term are correlated with the size of images. SemanticStyleGAN generates samples, which have resolution of 256×256 pixels, while DCGAN generates samples with a resolution of 64×64. In this way, learning features in higher resolution benefits the process. In addition, SemanticStyleGAN requires more iterations during the inpainting process, since the changes in the input vector cause only slight mutation in the output. This behavior is expected because the StyleGAN architecture is built to support such manipulations.

5   Discussion of Results

The qualitative and quantitative evaluation of the proposed SemanticStyleGAN model for inpainting images was performed on the FFHQ dataset. Both evaluations shed light on the capabilities and the limitations of the proposed model for inpainting.

5.1   Qualitative Results

The main objective of the models for inpainting images is to fill out a corrupted or missing region in an image with pixels perceptually and semantically consistent with the remaining parts of the image. Figure 3 shows a set of selected reconstructed images generated by the baseline DCGAN model [20], while Fig. 4 shows the outputs generated by the novel SemanticStyleGAN proposed in this research. These figures represent some of the best samples recovered by each of the models, and it can be noted that both models produce comparative results in terms of alignment and learning representation that is close to the original image. The two models differ on a more fine-grain scale, the reconstructed images generated by the SemanticStyleGAN contain fine and sharp details, compared to the blurrier and less convincing facial features generated by the baseline model DCGAN. StyleGAN is a generative model used to produce high-quality images and while the better quality of the generated images was expected, it proves that our choice of strengthening the model with style-based generator leads to inpainting images with superior quality. The quality of the images varies with the type of a mask (a shape of the corrupted region) being used. Regular large regions of missing pixels tend to be reconstructed with a great quality, since the generator produce samples that are consistent and realistic as a whole. Considering the manner in which the importance weights are computed, it was not surprising that masks covering a small area lead to generation of artifacts. Images corrupted by noise were recovered near perfectly in terms of shape and color, however, the presence of artificial artefacts poses a problem. When confronted with irregular, hand-drawn-stroke-like masks, we were faced with the appearance of blurred lines or reconstructed regions that do not match the image at all. One possible speculation for the latter results lies in the fact that the hand-drawn strokes cover various locations and objects of the image, hence the generative model was unable to satisfy all these requirements by matching them with appropriate importance term values. Additionally, the inpainting process was required to do many iterations before producing satisfying results.

We have also noticed examples of severely misaligned edges, or generated objects in places that they were not supposed to be, as shown on Fig. 5 and 6. Although we could assume that more iterations could improve the problem, the time and computational complexity might be impractical.

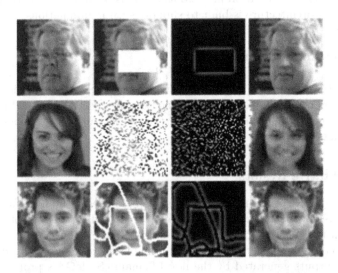

Fig. 3. Baseline DCGAN semantic inpainting model based on [20]. (Left to Right) Original image, input image, importance term and reconstructed image.

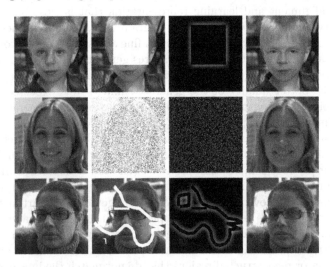

Fig. 4. SemanticStyleGAN inpainting model. (Left to Right) Original image, input image, importance term and reconstructed image.

Fig. 5. Problems and challenges of the baseline DCGAN semantic inpainting model based on [20]. (Left to Right) Original image, input image, importance term and reconstructed image.

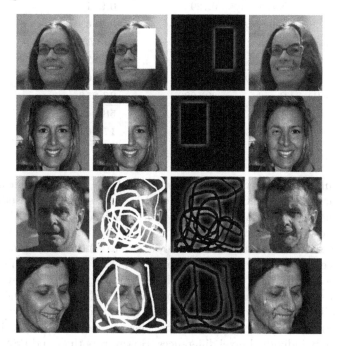

Fig. 6. Problems and challenges of the SemanticStyleGAN inpainting model. (Left to Right) Original image, input image, importance term and reconstructed image.

5.2   Quantitative Results

There is an ongoing discussion on whether the quantitative evaluation metrics show a clear view of the performance of the models for image inpainting, and yet they are frequently used for comparative analysis. A number of evaluation metrics have been chosen to measure the performance gains yielded by the proposed model, namely: ℓ_1, structural similarity index (SSIM), with a window size of 11, and a peak signal-to-noise ratio (PSNR). These metrics perform pixel-wise comparison, without any consideration of the perceptive quality of the generated images, hense, Fréchet Inception Distance (FID) [8] is used to measure the Wasserstein distance between embedded image features obtained from a pre-trained Inception-V3 model [19]. The performance results obtained by the baseline DCGAN model and our SemanticStyleGAN are presented in Table 1.

Table 1. Quantitative results over FFHQ dataset obtained by the baseline DCGAN [20] model and our SemanticStyleGAN model. The best result in each row is boldfaced.

Mask		Baseline DCGAN model	SemanticStyleGAN
ℓ_1 (%)	Regular	**2.25**	2.26
	Noise	5.6	**1.84**
	Quick Draw	**0.65**	0.7
SSIM	Regular	0.875	**0.878**
	Noise	0.729	**0.951**
	Quick Draw	0.964	**0.966**
PSNR	Regular	**25.481**	25.172
	Noise	20.745	**29.878**
	Quick Draw	34.055	**34.508**
FID	Regular	30.598	**28.856**
	Noise	203.981	**58.647**
	Quick Draw	25.351	**17.961**

The results indicate that the SemanticStyleGAN model significantly outperforms the baseline DCGAN based model in most scenarios. Baseline model showed slightly better results on ℓ_1 (%) error, which computes the absolute difference between the true and predicted pixel values of the images, which means that pixel-wise differences are smaller. However, ℓ_1 (%) captures even the slightest difference in pixel intensities and colors, and can be greatly influenced by a wide range of variations between pixels. SemanticStyleGAN achieves slightly better results over the baseline model in terms of PSNR; lower values indicate better overall pixel-wise reconstruction of the corrupted image. Unlike ℓ_1, PSNR is a quality measure independent of variations in image content, which is not overpowered by individual pixel differences. However, both ℓ_1 and PSNR assume pixel-wise independence, which may result in assigning favorable scores to perceptually inaccurate results. In contrast, SSIM performs image comparison based on luminance, contrast and structural information, which play an important role in capturing perceptual differences between images. Values for SSIM range from -1 to 1 and higher values mean greater similarity between the two images. From the results it can be seen that our proposed model obtains images with better perceptual quality when compared to the baseline model. It should be noted that all of the previously mentioned metrics measure the quality of the reconstructed image with respect to the ground truth, which is not the best measure since the generated image is not trying to reconstruct the ground truth image. In this regard, Fréchet Inception Distance measures the distance between the distribution of vision-relevant features from reconstructed and original images. As a result, FID does not focus on individual pixels nor structural information in the original image, but it tries to mimic human perception capability by making an educated comparison with what the target image features should look like. The FID scores of our proposed SemanticStyleGAN model are considerably lower than the baseline model, quantitatively showing that the images reconstructed using SemanticStyleGAN have a much better perceptual quality.

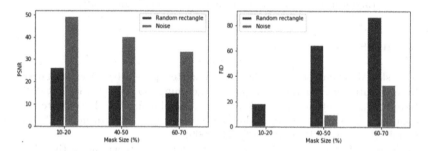

Fig. 7. Effects of mask shapes and sizes on PSNR and FID results obtained by our SemanticStyleGAN model.

Figure 7 gives an insight into the performance of our model for different shapes and sizes of masks. It can be noted that for larger corrupted regions, the quality of reconstruction deteriorates. Additionally, dispersed corruptions with higher amount of information per pixel obtained by noise masks, leads to a higher qualitative reconstruction quality, compared to corruptions that spread across entire region (missing rectangle) of the image.

6   Summary and Future Directions

In this paper an approach for image inpainting that exploits a style-based generator technique is proposed. The experiments show that the proposed GAN-based network achieves robust performance that is competitive and in some cases superior to previous attempts in the field. In particular, the experimental results has demonstrated that the model satisfies both, the qualitative and quantitative criteria for successful reproduction of a missing region in an image. Depending on the size and the shape of the missing region, the performance varies; larger area were almost perfectly reconstructed, while corruptions of small regions, spanning large distances, tend to face the problem of artifacts generation.

After an in-depth analysis of the results produced by the novel Semantic-StyleGAN model for image inpainting proposed in this paper, we highlight few directions for further extension of our model. Learning representation using pre-trained deep neural networks, such as VGG could mitigate the problem of replacing missing regions with pixels that do not fit perfectly with the surrounding areas. These feature vector representations would take into consideration individual features, instead of concentrating on a window around the corrupted region. One idea that could mitigate the problems related to slow inference is a technique of "hallucinating intermediate results" as proposed by the EdgeConnect model [14]. The two-step process of edge generation and image completion might speed up the reconstruction process, and produce images with finer details and quality in cases where the current version of the model faces challenges.

Acknowledgement. This work was partially financed by the Faculty of Computer Science and Engineering at the "Ss. Cyril and Methodius" University.

References

1. Arjovsky, M., Chintala, S., Bottou, L.: Wasserstein generative adversarial networks. In: International Conference on Machine Learning, pp. 214–223. PMLR (2017)
2. Barnes, C., Shechtman, E., Finkelstein, A., Goldman, D.B.: PatchMatch: a randomized correspondence algorithm for structural image editing. ACM Trans. Graph. (Proc. SIGGRAPH) **28**(3), 24 (2009)
3. Bertalmio, M., Sapiro, G., Caselles, V., Ballester, C.: Image inpainting. In: SIGGRAPH 2000, pp. 417–424. ACM Press/Addison-Wesley Publishing Co., USA (2000). https://doi.org/10.1145/344779.344972
4. Efros, A.A., Freeman, W.T.: Image quilting for texture synthesis and transfer. In: Proceedings of SIGGRAPH 2001, pp. 341–346 (2001)
5. Esedoglu, S.: Digital inpainting based on the Mumford-Shah-Euler image model. Eur. J. Appl. Math. **13**(4), 353–370 (2003). https://doi.org/10.1017/S0956792502004904
6. Goodfellow, I.J., et al.: Generative adversarial nets. In: Proceedings of the 27th International Conference on Neural Information Processing Systems, vol. 2, pp. 2672–2680. MIT Press, Cambridge (2014)
7. Hays, J., Efros, A.A.: Scene completion using millions of photographs. ACM Trans. Graph. (SIGGRAPH 2007) **26**(3) (2007)
8. Heusel, M., Ramsauer, H., Unterthiner, T., Nessler, B., Hochreiter, S.: GANs trained by a two time-scale update rule converge to a local nash equilibrium. In: Advances in Neural Information Processing Systems, pp. 6629–6640. Curran Associates Inc., Red Hook (2017)
9. Iskakov, K.: Semi-parametric image inpainting. arXiv preprint arXiv:1807.02855 (2018). https://arxiv.org/abs/1807.02855

10. Isola, P., Zhu, J.Y., Zhou, T., Efros, A.A.: Image-to-image translation with conditional adversarial networks. In: Proceedings of the IEEE Conference on Computer Vision and Pattern Recognition, pp. 1125–1134 (2017)
11. Johnson, J., Alahi, A., Fei-Fei, L.: Perceptual losses for real-time style transfer and super-resolution. In: Leibe, B., Matas, J., Sebe, N., Welling, M. (eds.) ECCV 2016. LNCS, vol. 9906, pp. 694–711. Springer, Cham (2016). https://doi.org/10. 1007/978-3-319-46475-6_43
12. Karras, T., Laine, S., Aila, T.: A style-based generator architecture for generative adversarial networks. In: Proceedings of the IEEE/CVF Conference on Computer Vision and Pattern Recognition, pp. 4401–4410 (2019)
13. Krizhevsky, A., Sutskever, I., Hinton, G.E.: Imagenet classification with deep convolutional neural networks. Adv. Neural. Inf. Process. Syst. **25**, 1097–1105 (2012)
14. Nazeri, K., Ng, E., Joseph, T., Qureshi, F., Ebrahimi, M.: EdgeConnect: structure guided image inpainting using edge prediction. In: The IEEE International Conference on Computer Vision (ICCV) Workshops, October 2019
15. Pathak, D., Krahenbuhl, P., Donahue, J., Darrell, T., Efros, A.A.: Context encoders: feature learning by inpainting. In: Proceedings of the IEEE Conference on Computer Vision and Pattern Recognition, pp. 2536–2544 (2016)
16. Pérez, P., Gangnet, M., Blake, A.: Poisson image editing. ACM Trans. Graph. **22**(3), 313–318 (2003). https://doi.org/10.1145/882262.882269
17. Radford, A., Metz, L., Chintala, S.: Unsupervised representation learning with deep convolutional generative adversarial networks. In: Bengio, Y., LeCun, Y. (eds.) 4th International Conference on Learning Representations, ICLR 2016, San Juan, Puerto Rico, 2–4 May 2016, Conference Track Proceedings (2016). http://arxiv. org/abs/1511.06434
18. Simonyan, K., Zisserman, A.: Very deep convolutional networks for large-scale image recognition. In: Bengio, Y., LeCun, Y. (eds.) 3rd International Conference on Learning Representations, ICLR 2015, San Diego, CA, USA, 7–9 May 2015, Conference Track Proceedings (2015). http://arxiv.org/abs/1409.1556
19. Szegedy, C., Vanhoucke, V., Ioffe, S., Shlens, J., Wojna, Z.: Rethinking the inception architecture for computer vision. In: Proceedings of the IEEE Conference on Computer Vision and Pattern Recognition, pp. 2818–2826 (2016)
20. Yeh, R.A., Chen, C., Yian Lim, T., Schwing, A.G., Hasegawa-Johnson, M., Do, M.N.: Semantic image inpainting with deep generative models. In: Proceedings of the IEEE Conference on Computer Vision and Pattern Recognition, pp. 5485–5493 (2017)
21. Yu, J., Lin, Z., Yang, J., Shen, X., Lu, X., Huang, T.S.: Generative image inpainting with contextual attention. In: Proceedings of the IEEE Conference on Computer Vision and Pattern Recognition, pp. 5505–5514 (2018)

NLP and Social Network Analysis

Models for Detecting Frauds in Medical Insurance

Hristina Mitrova and Ana Madevska Bogdanova[✉]

Ss. Cyril and Methodius University in Skopje, Rugjer Boshkovikj, 16, P. O. 393, Skopje 1000, North Macedonia
`hristina.mitrova@students.finki.ukim.mk,`
`ana.madevska.bogdanova@finki.ukim.mk`

Abstract. Health insurance is important for many people, but unfortunately it is susceptible to frauds, therefore expenditures for covering the funds show exponential growth. The victims of this kind of scams are not only the institutions that provide the funds and treatments, but also are the ones who really need that help, except they have lost their priority due to a committed fraud. In order to rationally provide funds and minimize losses, there is a need for fraud detection systems. In this paper, this issue is considered as a binary classification problem, using data inherent in the nature of the field. The whole data science pipeline process is considered in order to elaborate our results that are higher than the published ones on the same problem: 0.95, 0.96 and 0.98 AUC scores with different models. The data is integrated from three interconnected databases, which are pre-processed and then their cross-section is undertaken. The dataset is unbalanced concerning the records of both classes, therefore certain balancing techniques are applied. Several models are built using traditional Machine Learning models, classifiers with Deep Neural Networks and ensemble algorithms and their performance is validated according to several evaluation metrics.

Keywords: Health insurance · Detection · Binary classification · Fraud · Datasets · Preprocessing · Algorithms · Ensemble methods · Balancing techniques · Evaluation metrics · Hyperparameter tuning

1 Introduction

Many analyzes related to health insurance cases have concluded that different stakeholders are involved in the fraud schemes, including the healthcare staff. They ambiguously present health diagnoses in order to receive compensation that corresponds to high-paying drugs, medical procedures and treatments. Health insurance scams can take many forms, and some of the most common are:

- recovery for services that have not been used
- double application for the same service
- misrepresentation of service

L. Antovski and G. Armenski (Eds.): ICT Innovations 2021, CCIS 1521, pp. 55–67, 2022.
https://doi.org/10.1007/978-3-031-04206-5_5

- payment for a service that is more complex and more expensive than the one that is actually made
- reimbursement for covered service, when the one provided is not covered.

In order to prevent these illegal activities, the insurance beneficiaries are responsible to follow the process and if they notice a certain error or inconsistency, they are required to report to their insurance company [1]. The insurance company itself is obliged to investigate the authenticity of the case, but this would sometimes require consultation with a number of institutions, as well as the submission of the patient's complete history documentation as evidence of his/her need for treatment funds. In reality, this problem can be very complex and there is a need to implement software solutions that are able to detect fraud by using automated data processing [2]. In this paper, we discuss the detection of frauds as a binary classification problem by using different Machine learning methods for building the models - traditional Machine learning methods, deep learning methods and some ensemble algorithms. The solution uses three content connected datasets, separately preprocessed and integrated into one dataset. Preprocessing is one of the most essential parts in data science that leads to more successful model performance. Despite the appropriate handling of data noise, another important piece of the puzzle is a good data exploratory analysis that helps with the interpretability of the issue at hand. Frequently used techniques for imbalanced data include data sampling, in order to achieve smaller differences between the class occurrence. We concluded that the degree of the class balance has a valuable impact on model performance and setting the appropriate parameters is an important factor in the higher validation score of a given model.

The rest of the paper is organized as follows. The second section elaborates related work, the next section describes the methods used to build and evaluate our models. The last section includes some important conclusions and comparison with the published results on the same problem.

2   Related Work

In this section, we provide information about some public research associated with our subject of interest.

Authors in [3] use a similar approach in preparing the data set by integrating multiple data sets into one, using the following classification algorithms: Logistic Regression, Random Forest Classifier with 100 base estimators and Gradient Boosting Tree. Publicly available are only results evaluated by AUC score. The data integration is done using Big Data datasets of different parts of the Medicare program from Centers for Medicare and Medicaid Services (CMS) - "Medicare Provider Utilization and Payment Data":

- Part D data set that provides information on prescription drugs that are included in the program offered by medical care on an annual basis;
- Part B data set that provides information on each medical intervention performed by a healthcare professional on an annual basis;
- DMEPOS database that provides information on medical equipment, prosthetics, orthotics and similar helping aids.

The following study [4] used Neural Network-based models to detect fraud, using the same data processed in our paper provided by the same health program, with the difference that the data in this specific cited study covers the period from 2012 to 2016. The data provides information for medical interventions, used services, medical staff involved and a list of excluded people and entities (LEIE - List of Excluded Individuals and Entities), used for data labeling. The evaluation of the classifiers is utilizing ROC AUC, TPR, TNR and G-Mean results. Classifiers such as Logistics Regression, Gradient Boosting Trees and Neural Networks implemented in Keras on the TensorFlow platform are used. The Neural Network method stands out as the most successful among them.

Research paper [5] integrates data published by (LEIE) that represents information in the period between 2012 and 2015 calendar years. They deal with unbalanced data with the undersampling technique, and the performance metrics used for evaluation are AUC score, FPR and FNR, while the models are based on methods such as Decision Trees, Logistic Regression and Support Vector Machine classifier.

In addition to building models with Machine Learning methods, the topic of fraud detection is also considered from a financial point of view. The purpose of the research study [6] is to explore the current state of Medicare fraud in the U.S., identify current policies and laws that foster Medicare fraud, and to determine the financial impact of Medicare fraud. The methodology for this study is a literature review, conducted using a scholarly online database search and government websites. The problem in research [7] is discussed including the payers, providers, and patients. The knowledge base, covered through the sections, is represented by fraud detection literature and the state of the industry. Based on this analysis, a multidimensional schema based on Medicaid data is developed, in order to describe a set of multidimensional models and techniques to detect fraud in large sets of claim transactions, later evaluated through functionally testing against known fraud schemes. Research [8], reviews studies that performed data mining techniques for detecting health care fraud and abuse, using supervised and unsupervised data mining approaches. Most available studies have focused on algorithmic data mining without an emphasis on or application to fraud detection efforts in the context of health service provision or health insurance policy.

3   Proposed Solution

In the following section, we elaborate the used approaches for developing the best models for the given fraud detection problem.

3.1   Methods for Building the Models

Finding the solution for the posed problem, several models were built using a variety of Machine Learning methods, including the ones with a traditional approach such as.

KNN, Decision Tree, Logistic Regression, Support Vector Machines and Gaussian Naive Bayes. The models built with these methods showed solid results. This plethora of methods was used in order to compare our results with the different models published in [3–5].

In addition, a model based on Deep Neural Networks was developed - three hidden layers based on the Multi Layer Perceptron classifier. As an activation function that calculates complex patterns, ReLU was used, since it does not activate all neurons at the same time and thus provides high computational efficiency.

As a third approach, ensemble algorithms were developed, as methods for building models that include a group of weak models that combine to form a strong learning model. We used ensemble methods such as Bootstrap Aggregation (bagging) and Boosting, and as their base estimators we used the approach with Support Vector Machine and Decision Trees. These two types of algorithms are different [9] since

- bagging algorithms aim to reduce the variance of the decision tree, creating several sets of data from a randomly selected training sample selected by substitution;
- boosting algorithms, an ensemble technique is applied to create a collection of classifiers where the models learn sequentially, with each subsequent model learning from the errors of the previous models through data analysis.

When a record is incorrectly classified by a hypothesis, weights are added to the next model that increase the likelihood that the next hypothesis will be correctly classified. By combining such weak models, the outcome are highly successful models.

3.2 Evaluation Metrics

The experiments were evaluated utilizing several metrics, based on specific types of error such as accuracy, precision, recall and f1 score and also ROC with AUC score were displayed. Additionally, False positive rate (FPR) and False negative rate (FNR) were extracted from the appropriate confusion matrix when these parameters were needed to compare our results with the published ones. According to the fact that the data is imbalanced, not only that we were interested in the results based on the number of accurately predicted classes represented by the accuracy metric, but also the percentage of accurately predicted records of all records that are positive or how many of the positively predicted records are in fact positive, which gives us the information about the model's precision. Another important matter is the model's sensitivity that represents the percentage of accurately predicted records of all records that are certain to belong to the target class.

4 Experiments and Results

In this section, we give a detailed description of the whole pipeline: preprocessing and data exploration, models that are built and the evaluation metrics used in the experiments. The data preprocessing and exploratory analysis was performed using Python [10], along with some libraries including Pandas, Numpy, Matplotlib and Imblearn [11–14]. The models were built using method implementations from Sklearn [15].

4.1 Dataset

The data was collected as part of a health program called Medicare Provider in the United States to analyze the construction of a software system that could detect fraud and is publicly available on Kaggle [16]. We work with a set of information that is combined from three data sets:

- A database that provides details about clients' health status, region in which they live and other personal information;
- A database that provides information on hospital treatment, providing insight into claims submitted to those patients admitted to hospitals with additional details such as their date of admission and discharge and the code for receiving the diagnosis;
- A data set that provides outpatient data with details of those patients who visit hospitals but are not admitted there.

4.2 Preprocessing and Data Exploratory Analysis

One of the most essential parts in data science is the preprocessing part. In order to build models with appropriate degree of confidence, the data missing values were handled.

Handling the Missing Values. In the first data set, the only missing values were in the DOD-Date of death attribute. For the records with no DOD value, it is considered that the patients are still alive, therefore a new attribute was extracted which indicates the age (Age). After this intervention, there were no more missing values in the first data set. The second and the third dataset lacked values on characteristics related to financial assets. Due to the appearance of outliers, it was necessary to take the median as a value that will be filled to the missing parts, since the average is sensitive to outliers and it would contribute to misrepresentation of the data. For the attributes that were intended to indicate that a certain person from the health staff participated in a medical treatment, an indexed identification value was assigned. As there were records with no information about this feature, it was filled with 0 value and the records with an identification were replaced with the value of 1.

Data Integration. Following the separate preprocessing of the data sets, the common features from all three datasets were selected (33 features). In order to merge the data, we assigned the feature "Insurance provider" as a key that cohorts the data (same beneficiary/patient used the insurance multifold under the same provider). In this kind of grouping for the features that represent monetary compensation or the number of medical personnel involved in the treatments of the same beneficiary, a cumulative value was taken. Finally, labels were added to this set, which were also connected through the appropriate Insurance provider. Entries with the label "yes" were replaced with the value of 1, while those with a "no" were replaced with the value of 0.

Data Normalization. After the data integration, a copy was saved with applied MinMax normalization because some of the algorithms for model building require normalized data.

Data Exploratory Analysis. As a next step, a Data exploratory analysis was performed in order to enhance the understanding of the data set - the degree of correlation among the features (attributes) and direct influence of every feature to the Fraud outcome, through correlation matrix. Those coordinates that were colored in a darker color represented a higher degree of correlation (Fig. 1). As an example, it can be noticed that the existence of a certain chronic disease, the involvement of a medical person, the amount of money that is confirmed to be covered by insurance, as well as the days for inpatient admission are correlated with fraudulent outcomes. This qualitative assessment may help to improve the interpretability of the classification results.

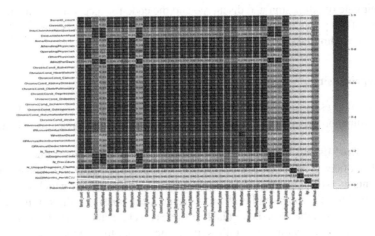

Fig. 1. Feature correlation matrix

Solving Data Imbalance. The records in the consolidated data set, labeled as potential fraud (positive class) are 506 in total, while those labeled as regular (negative class) are 4904, i.e. 90.65% of the data belong to the negative class and the remaining 9.35% to the positive class. The data set is highly unbalanced in terms of the two classes. In order to emphasize the need of balancing the classes, we constructed series of models built on data without taking into consideration that there is data imbalance, and latter we applied cost sensitive classification and decided to use some sampling techniques such as Random Undersampling with 80% balance, Random Oversampling with 75% balance, combination of oversampling and undersampling with 75% balance and oversampling with 100% balance using SMOTE.

4.3 Building Models and Evaluation

The data set is randomly divided into 70% for training and 30% of the records are set aside for testing sets. In order to find the most successful model, the data set is trained in 6 different ways including training data without taking into account that the data is imbalanced, dealing with data imbalance by applying cost-sensitive classification and later different sampling techniques with various levels of balance as mentioned in the previous section. In this section we are providing detailed results from the classification techniques that stood out as the most successful:

- Training with imbalanced data;
- Cost sensitive classification on imbalanced data;
- SMOTE for reaching full level of balance.

In order to find the best model and to compare the validation values with the already published results on the same problem, we utilized the following metrics: accuracy, precision, recall, f1 score, AUC score.

Training with Imbalanced Classes. Models using the following algorithms were used in this section: Decision Tree Classifier, Logistic Regression Classifier, Random Forest and Support Vector Machine Classifier for classification (Table 1).

Table 1. Results of the first type of classification

Metric\Model	Decision Tree Classifier	Logistic Regression	Random Forest Classifier	SVM
Accuracy	0.9067	0.7062	0.9261	0.9122
Precision	0.5269	0.1832	0.6806	0.5714
Recall	0.4623	0.5755	0.4623	0.4151
F1 score	0.4925	0.2779	0.5506	0.4809
AUC score	0.75	0.61	0.91	0.75

The Random Forest-based classifier stands out as the most successful model, but due to unsatisfactory results for other metrics besides accuracy, some techniques to deal with class imbalances that affect our models' performance were used.

Cost Sensitive Classification. In this part, the models are trained without applying a balancing technique, but certain improvements occur whenever the model is penalized for incorrect prediction. Adding weights is done by setting the class weight parameter to receive a value for balance (class_weight = 'balanced'). Most Machine Learning models are trained to minimize error, adding weight to the value of the misclassification error determined by the proportionality of the class (Table 2).

Table 2. Results from the cost sensitive classification

Metric\Model	Decision Tree Classifier	Logistic Regression	Random Forest Classifier	SVM
Accuracy	0.9104	0.8909	0.9131	0.8667
Precision	0.5364	0.4647	0.5469	0.3911
Recall	0.6419	0.7453	0.6604	0.8302
F1 score	0.5837	0.5725	0.5983	0.5317
AUC score	0.80	0.92	0.91	0.91

It can be noted that the cost-sensitive classification shows significant improvements in the recall and some improvements in other metrics, as opposed to the previous training approach.

Classification After Applying SMOTE Technique. SMOTE oversampling (Synthetic Minority Oversampling Technique) is a sampling technique for adding records in the minor class in order to completely equalize the frequency of records from both classes. This algorithm helps to overcome the overfitting problem posed by random over-sampling. It focuses on the feature space to generate new instances with the help of interpolation between the positive instances that lie together. According to this technique, the representation of the classes was 4904 records in the positive and 4904 records in the negative class, which further increased the number of data (Table 3).

Table 3. Results after applying SMOTE balancing technique

Model/Metric	Decision Tree	SVM	KNN	MLP	Random Forest	Bagged Decision Tree	ADA	Gradient Boos ting Tree	XG Boost	Logistic Regression
Accuracy	0.93	0.88	0.92	0.88	0.94	0.94	0.95	0.92	**0.96**	0.86
Precision	0.91	0.86	0.89	0.86	0.92	0.91	0.94	0.89	**0.94**	0.89
Recall	0.95	0.98	0.96	0.92	0.98	0.97	0.97	0.94	0.97	0.82
F1 score	0.93	0.88	0.93	0.88	0.95	0.94	0.95	0.92	**0.96**	0.85
AUC score	0.93	0.94	0.92	0.95	0.99	0.98	0.98	0.98	**0.99**	0.95

The full balance of the data with SMOTE benefits in favor of significant performance improvements in the models in relation to all evaluation metrics for classification. The models that use ensemble algorithms stand out as the most successful, as a result of the

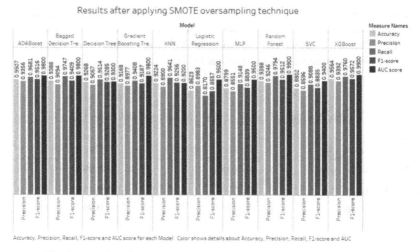

Fig. 2. Bar plot with results after applying SMOTE balancing technique

large number of interpreters used in reasoning. The most successful result was obtained by applying the Extreme Gradient Boost (XGBoost) classifier with Decision Tree as the base estimator, with adjusted hyperparameters in order to achieve the best performance measured according to the evaluation metrics. In this setting, we concluded that when the number of estimators is 110 and the maximum branching depth is 10, better results for the evaluation metrics are achieved.

4.4 Best Results

Random Forest proved to be the most successful in training the models on unbalanced data, and that being a type of ensemble method resulted with motivation to apply other ensemble methods for building new models. These models surpassed the results of models based on previous methods and proved to be the most successful when SMOTE technique was applied. XGBoost stood out as the most effective, followed by the model based on Adaptive Boost. In order to be sure of the success of the best models, we made 5 fold and 10 fold cross validation and concluded that the obtained results match the previously obtained ones and that there are no meaningful differences between them.

5 Comparison with Closely Related Research Papers

In this section a comparison is made between the published results solving the insurance fraud problem with ML, elaborated in Sect. 2. Related Work and our best results obtained from the classification models in Sect. 4.3, **Classification after applying SMOTE technique** (noted as **'Our results'** in the following Tables), according to the corresponding methods and evaluation metrics used in each of the papers, accordingly.

- The authors in the paper [3] use integration of three data sets provided by the same medical program [2], with 90% resemblance with our data set. Their results are evaluated with the AUC score. The following Table 4 gives comparison with the results obtained with our research, in favor of our best model.

Table 4. Comparing our AUC scores with research paper [3]

Method\AUC score	Research paper [3]	Our results
Random Forest Classifier	0.79383	**0.99**
Gradient Boosting Tree	0.79047	**0.98**
Logistic Regression	0.81554	**0.95**

- Paper [4] builds models using methods based on Deep Neural Networks (DNN), utilizing data provided by the same medical program. Due to the approaches in data preprocessing, there is an important amount of resemblance with our data that is corresponding, thus the nature of the data represents equivalent information. The obtained results are in favour of our models (Table 5).

Table 5. Comparing our AUC scores, TPR and TNR with research paper [4]

DNN\Evaluation metric	Research paper [4]	Our results
AUC score	0.8058	**0.9500**
Recall – TPR (True Positive Rate)	0.8280	**0.9148**
Specificity – TNR (True Negative Rate)	0.6099	**0.8449**

- Research [5] uses models based on Logistic Regression, Decision Tree Classifier and Support Vector Machine classifier. It should be noted that we expect AUC scores to have higher values and lower values for FPR and FNR in order to confirm the model's good performances. The first three rows represent the results from the aforementioned research, followed by the results attained with our models in this paper. AUC score is in favour of our models (Table 6).

There is always a need to aim for results improvement, taking into account the reliability of the model. Compared to the elaborated papers in this Section, our model differs in taking into consideration missing data, class imbalance and applying hyperparameter tuning.

Table 6. Comparison of results by AUC score, FPR and FNR with research paper [5]

ML Method	Research paper	AUC score	FPR	FNR
Logistic Regression	Research paper [5]	0.880	0.099	0.411
Decision Tree	Research paper [5]	0.882	0.191	0.226
SVM	Research paper [5]	0.861	0.102	0.416
Logistic Regression	Our results	0.95	0.70361	0.183
Decision Tree	**Our results**	**0.93**	**0.0978**	**0.0486**
SVM	Our results	0.94	0.1484	0.0912

6 Conclusion

In this paper a thorough investigation is done over the health insurance fraud detection problem, solved by building several Machine Learning models. Our goal was to explain the whole pipeline Data science process in detail, in order to elaborate the obtained results that are better than several published ones [3–5].

Following the undertaken study, we came to some interesting conclusions that were related to different aspects in each of the training sections. The degree of equilibrium in sense of class balance always plays a big role and its increase has a direct proportion to the increase of the evaluation metric results, but when choosing a sampling technique with the same degree of balance, the models built on data sampled with oversampling technique are more successful. The different proportion of instances in the training and testing data sets (70/30, 80/20) does not give significant differences in performance, but it is very important to properly tune the hyperparameters. From an algorithmic point of view, the models whose classifiers are based on ensemble algorithms have proved to be the most successful for this problem, due to their mechanism of joint decision-making by multiple internal classifiers, which study different subsets of the training data set. The model based on the XGBoost method gave the highest results, F1 score of 0.96 and AUC score of 0.99.

As a motivation for further improvement of the built models, an approach of comprehensive tunning of the hyperparameters of the algorithms can be applied, in order to reach even more successful models, while the analysis of the correlation of the attributes can help explain the obtained results. A future aspect of model improvement would include analyzing the impact of attributes in the data set and determining the degree of their importance in achieving a better result. A commonly used technique for this approach is to compare attributes through the Shapley value [18], which determines the importance of each of the attributes. This value would help in more optimal separation and grouping of attributes in order to build models that would improve performance. By the undertaken exploratory analysis it is noticed that the existence of a certain chronic disease, the involvement of a medical person, the days for inpatient admission, as well as the amount of money that is confirmed to be covered by insurance are correlated with the dependent variable that indicates whether a particular instance is fraudulent or not. This qualitative assessment may help to explain some of the classification results.

Furthermore, exploratory analysis leads to better interpretability of the system, distinguishing the feature importance for a given model to provide a group of features most correlated to the class label's value.

References

1. Medicare.gov. https://www.medicare.gov/forms-help-resources/help-fight-medicare-fraud/how-spot-medicare-fraud. Accessed 15 Feb 2021
2. Fraud Prevention System - Department of Health & Human Services, Centers for Medicare & Medicaid Services. https://www.cms.gov/About-CMS/Components/CPI/Widgets/Fraud_Prevention_System_2ndYear.pdf. Accessed 15 Feb 2021
3. Herland, M., Khoshgoftaar, T.M., Bauder, R.A.: Big data fraud detection using multiple medicare data sources. J. Big Data **5**, 29 (2018). https://doi.org/10.1186/s40537-018-0138-3. Accessed 23 Mar 2021
4. Johnson, J.M., Khoshgoftaar, T.M.: Medicare fraud detection using neural networks, J Big Data **6**, 63 (2019). https://doi.org/10.1186/s40537-019-0225-0. Accessed 23 Mar 2021
5. Bauder, R.A., Khoshgoftaar, T.M.: The detection of medicare fraud using machine learning methods with excluded provider labels, science direct. In: The Thirty-First International Florida Artificial Intelligence Research Society Conference (FLAIRS-31) (2018). https://www.sciencedirect.com/science/article/pii/S2212017313002946?via%3Dihu. Accessed 23 Mar 2021
6. Hill, C., Hunter, A., Johnson, L., Coustasse, A.: Medicare Fraud in the United States: Can it ever be stopped? Marshall University, Marshal Digital Scholar (2014). https://core.ac.uk/download/pdf/232719599.pdf. Accessed 24 Mar 2021
7. Thornton, D., Mueller, R.M., Schoutsen, P., van Hillegersberg, J.: Predicting healthcare fraud in medicaid: a multidimensional data model and analysis techniques for fraud detection. In: CENTERIS 2013 - Conference on Enterprise Information Systems/ProjMAN 2013-International Conference on Project Management/HCIST 2013 - International Conference on Health and Social Care Information Systems and Technologies. https://www.researchgate.net/publication/259576210_Predicting_Healthcare_Fraud_in_Medicaid_A_Multidimensional_Data_Model_and_Analysis_Techniques_for_Fraud_Detection. Accessed 24 Mar 2021
8. Joudaki, H., et al.: Using data mining to detect health care fraud and abuse: a review of literature. Glob. J. Health Sci. **7**(1) (2015). ISSN 1916-9736 E-ISSN 1916-9744. Published by Canadian Center of Science and Education. https://www.researchgate.net/publication/270652005_Using_Data_Mining_to_Detect_Health_Care_Fraud_and_Abuse_A_Review_of_Literature. Accessed 24 Mar 2021
9. González, S., García, S., Del Ser, J., Rokach, L., Herrera, F.: A practical tutorial on bagging and boosting based ensembles for machine learning: algorithms, software tools, performance study, practical perspectives and opportunities. Inf. Fusion **64** (2020)
10. Python: A programming language. https://www.python.org/. Accessed 11 Feb 2021
11. Pandas: An open source python library. https://pandas.pydata.org/. Accessed 11 Feb 2021
12. NumPy: An open source python library. https://numpy.org/. Accessed 11 Feb 2021
13. Matplotlib: An open source python library. https://matplotlib.org/. Accessed 11 Feb 2021
14. Imblearn: an open source python library. https://imbalanced-learn.org/stable/. Accessed 11 Feb 2021
15. Sklearn, an open source python library. https://scikit-learn.org/stable/. Accessed 11 Feb 2021
16. Publicly available data on Kaggle, Healthcare provider fraud detection analysis. https://www.kaggle.com/rohitrox/healthcare-provider-fraud-detection-analysis. Accessed 10 Feb 2021

17. Chen, T., Guestrin, C.: XGBoost: A Scalable Tree Boosting System, University of Washington, https://arxiv.org/pdf/1603.02754.pdf. Accessed 01 Apr 2021
18. Shapley value. https://en.wikipedia.org/wiki/Shapley_value. Accessed 01 Apr 2021
19. Markus, J., Pinter, M., Raduvanyi, A.: The Shapley value for airport and irrigation games, Corvinus University of Budapest (2011). https://www.academia.edu/47921066/The_Shapley_value_for_airport_and_irrigation_games. Accessed 1 Apr 2021
20. Wang, R.: AdaBoost for feature selection, classification and its relation with SVM, a review. In: 2012 International Conference on Solid State Devices and Materials Science (2012)
21. Amin, M.Z., Ali, A.: Application of Multilayer Perceptron (MLP) for data mining in healthcare operations. In: 3rd Conference on Biotechnology, University of South Asia, Lahore, Pakistan (2017)

Multi-criteria Evaluation of Students' Performance Based on Hybrid AHP-Entropy Approach with TOPSIS, MOORA and WPM

Iliyan Petrov[✉]

Department of "Information Processes and Decision Support Systems", Bulgarian Academy of Sciences, Institute of Information and Communication Technologies, akad. Georgi Bonchev srt., bl. 2, 1113 Sofia, Bulgaria
liyan.petrov@iict.bas.bg

Abstract. Worldwide pandemics and long periods of social containment require further improvement of Intelligent Education System (IES) with reliable communication networks and computing systems for providing accessible and effective teaching and learning. One of the main challenges in distance learning is to ensure adequate monitoring of the education process and evaluation of interim and final results. The paper presents a hybrid approach for multi-criteria evaluation of students' performance in a flexible framework of three preference scenarios where the theoretical learning, practical skills, and final exams participate with different weights. The systematization of criteria in blocks and their weighting according to preferences allow obtaining more objective macroscopic results. The Multiple Assessment integration (MAI) of the evaluation values in TOPSIS, MOORA and WPM allows to explore their behavior with different data sets and contributes to consolidate the final results for obtaining a better holistic and personalized view of the education process for each individual student.

Keywords: AHP · Entropy · Intelligent Education System · MCDM · TOPSIS · MOORA · WPM

1 Introduction

IES are specific cyber-physical systems comprising computing, networking, and sensing technologies for realization and management of education processes [1–6] in which the assessment of students remains an integral part of the learning process in middle and high-level education programs and has to take into account different criteria and formats [7–9]. Multi-Criteria Decision Making (MCDM) is one of the main disciplines in operations analysis which employs different subjective (expert-dependent) and objective (data-driven) methods for assessing the quality and performance of complex systems in many areas [10–14]. The inconveniences of relying on experts' opinions are determined by the subjectivity of human judgements, while, on the other hand, the data-driven selections depend on the quality of data and often neglect the needs of users. In practice, one-sided subjective or analytical methods may lead to uncertainty in the decision-making process.

© Springer Nature Switzerland AG 2022

L. Antovski and G. Armenski (Eds.): ICT Innovations 2021, CCIS 1521, pp. 68–84, 2022.

https://doi.org/10.1007/978-3-031-04206-5_6

2 Objective Definition of Criteria Weights with Entropy

The Information Entropy concept allows performing an objective definition of weights for evaluation criteria [15]. It is used for different MCDM problems and includes several consecutive steps. In this paper, we use the essentials of common MCDM methodology aspects developed in our previous work [16].

Step 1. This step prepares the data for the selection process.

<u>Step 1.1.</u> Definition of the *Basic Decision Matrix (BDM)* comprising "*m*" target alternatives (A_i) which are evaluated with a set of "*n*" independent criteria (C_j):

$$BDM_{m \times n} = [x_{ij}]_{m \times n}, \text{ where } i = 1, 2, \ldots, m; j = 1, 2, \ldots, n \qquad (1)$$

In this study, the modelling and simulations are applied on set of 10 alternatives (students) which are assessed by 9 criteria including traditional and IES related parameters, such as time spent on quizzes, peer and self-assessment [17], shown in Table 1.

Table 1. Target alternatives and evaluation criteria for selection of RE projects

Evaluation Criteria			Criteria type	Blocks
C1	Absence from lectures	A	Non-Benefit/Cost	
C2	Attention & Participation	A&P	Benefit/Profit	Learning
C3	Theory Quiz	TQ	Benefit/Profit	
C4	Time on Theory Quiz	TTQ	Non-Benefit/Cost	
C5	Peer assessment (on practice assignment)	PA	Benefit/Profit	
C6	Self-Quiz (Practice)	SQ	Benefit/Profit	Practice
C7	Time on Self-Quiz	TSQ	Non-Benefit/Cost	
C8	Exam	E	Benefit/Profit	Exam
C9	Time on Exam	TE	Non-Benefit/Cost	

The Basic Decision Matrix (BDM) of this study is displayed in Table 2.

<u>Step 1.2.</u> Calculating the proportions of relative parts p_{ij} for each value of x_{ij} in C(i) and formation of the *Normalized Decision Matrix NDM* $= [p_{ij}]_{m \times n}$:

$$p_{ij} = x_{ij} / \sum_{i=1}^{m} x_{ij}, \text{ with } \sum_{i=1}^{m} p_{ij} = 1 \qquad (2)$$

Step 2. Non-linear normalization of p_{ij} to obtain "*normalized performance ratings r_{ij}*" in Eq. (3) and forming of Normalized Performance Matrix *(NPM $= [r_{ij}]_{m \times n}$)*:

$$r_{ij} = p_{ij} / \left(\sum_{i=1}^{m} p_{ij}^2 \right)^{-2} \qquad (3)$$

Table 2. Basic Decision Matrix (BDM)

Criteria (n)	C1	C2	C3	C4	C5	C6	C7	C8	C9
	A	A&P	TQ	TTQ	PA	SQ	TSQ	E	TE
A1	0,01	0,91	0,82	1	0,62	0,92	0,9	0,98	0,85
A2	0,05	0,94	0,84	0,95	0,72	0,62	1	0,96	0,9
A3	0,07	0,81	0,74	1	0,92	0,82	0,8	0,82	0,92
A4	0,15	0,84	0,83	0,95	0,91	0,63	0,8	0,55	1
A5	0,25	0,71	0,73	1	0,89	0,72	0,75	0,78	0,9
A6	0,15	0,44	0,59	1	0,61	0,74	1	0,67	1
A7	0,2	0,66	0,71	1	0,82	0,61	1	0,1	1
A8	0,35	0,42	0,72	1	0,52	0,63	1	0,41	1
A9	0,2	0,72	0,84	0,85	0,46	0,86	0,95	0,52	1
A10	0,01	0,91	0,73	0,9	0,62	0,95	0,8	0,95	0,95

Step 3. Traditional Shannon Entropy transformation of primary data.

Step 3.1. Calculation of individual Shannon Entropy values (se_{ij}) for each nominal *relative weight* p_{ij} in *NDM* computable in different logarithmic formats in Eq. (4):

$$se_{ij} = -p_{ji} \cdot \log_a(p_{ij}) \tag{4}$$

Step 3.2. Calculation of the cumulative nominal Shannon Entropy [18] for each criterion "j" as a sum of the "i" individual entropies for the "m" target alternatives:

$$SEnom_j = \sum_{i=1}^{m} se_j = -\sum_{i=1}^{m} p_{ij} \cdot \log_2(p_{ij}) \tag{5}$$

Step 3.3. Calculation of maximal Shannon Entropy value for a set of "m" alternatives in which all elements have equal relative weights, i.e. $p_{ij} = 1/m$:

$$SEmax(m) = -\sum_{i}^{m} (1/m \cdot \log_2 1/m) = \log_2 m \tag{6}$$

Step 3.4. Normalization of $SEnom_j$ by comparing it to $SEmax(m)$:

$$SEnorm_j = e_j = SEnom_j/SEmax(m) = -(\log_2 m)^{-1} \sum_{i=1}^{m} p_{ij} \cdot \log_2(p_{ij}) \tag{7}$$

Step 3.5. Calculation of the sum of the normalized criteria entropies $e_j = SEnorm_j$ and defining their "*real entropy criteria weights (recw_j)*" for each criterion C_j:

$$recw_j = e_j / \sum_{i=1}^{n} e_j) \tag{8}$$

Step 3.6. Calculation of "*differed entropy criteria weights (decw_j)*" based on additional data treatment whose "aim" is to expand the difference between criteria through a "divergence parameter" d_j which is an arithmetic difference between the normalized maximal entropy $SEmax(m)$ expressed as "1" minus $SEnorm_j =$ "e_j":

$$d_j = 1 - e_j \tag{9}$$

$$decw_j = d_j / \sum_{i=1}^{m} d_j = (1 - e_j) / \sum_{i=1}^{m} (1 - e_j) \tag{10}$$

Table 3. Entropy transformation. Real and differed weights in stand-alone Entropy approach

Criteria (n)	A	A&P	TQ	TTQ	PA	SQ	TSQ	E	TE
A1	0,050	0,373	0,348	0,339	0,307	0,371	0,332	0,404	0,311
...	...	...	...	...	...	...	...	...	...
A2	0,050	0,373	0,326	0,319	0,307	0,378	0,310	0,398	0,332
SEnom	2,885	3,277	3,315	3,320	3,285	3,303	3,313	3,180	3,320
SEmax	3,322	3,322	3,322	3,322	3,322	3,322	3,322	3,322	3,322
SEnorm=e_j	0,869	0,986	0,998	0,999	0,989	0,994	0,997	0,957	0,999
recw$_j$	0,099	0,112	0,114	0,114	0,113	0,113	0,113	0,109	0,114
$d_j=1-e_j$	0,131	0,014	0,002	0,001	0,011	0,006	0,003	0,043	0,001
decw$_j$	0,624	0,064	0,010	0,003	0,053	0,028	0,012	0,203	0,003

The results of Steps 3.1–6 are displayed in Table 3.

The real entropy criteria weights are deducted directly from the empiric data and on this basis should look quite objective. However, in some cases, this totally data-driven approach may seriously underestimate the most meaningful and important attributes. In our case, the weight of final exam "E" is only 0.109, and lower than the average of 0.111. At the same time, while the values of "$recw_j$" are quite similar, the values of "$decw_j$" become very unequal. As a result, only one interim education parameter ("Absence") concentrates 0,624 of the total weights, while the "Self Quiz" (0,028) and the "Theory Quiz" (0,02) are practically neglected. In this type of extractions, the losers in terms of diversity are transformed into winners in terms of concentration. In our opinion, the concept of "information extraction" from a single entropy value is a mathematical manipulation, and if there are doubts about the information entropy approach the logical way-around should be to measure the opposite of diversity. In other words, to assess directly the concentration of information in each criterion rather than trying to "extract some value" from entropy for judging indirectly about concentration. The substitution of "*real entropy criteria weights* ($recw_j$)" with another set of "*differed entropy criteria weights* ($decw_j$)" produces very different results without taking into account the preferences of users. In a theoretical discussion, any data treatment might be regarded as an experiment but in a real situation, such approximate reality may transform the positioning of all parameters. In both cases, these stand-alone approaches are very formal and abstract, while the main task in MCDM is to select the alternatives that correspond to the preferences of users.

Step 4. Calculation of "normalized weighted values (ratios) "v" with real weights ($r_{ij}.recw_j = rv_{ij}$) and with differed weighs ($r_{ij}.decw_j = dv_{ij}$) for constructing the Normalized Weighted Matrices (*NWM*):

$$NWM\,(R)_{mxn} = [r_{ij}.recw_j]_{mxn} = [rv_{ij}]_{mxn} \qquad (11)$$

$$NWM\,(D)_{mxn} = [r_{ij}.decw_j]_{mxn} = [dv_{ij}]_{mxn} \qquad (12)$$

Step 5. Final assessment and ranking of results with different evaluation techniques.

3   Front-End AHP: Systemization of Evaluation Criteria in Blocks

The stand-alone Entropy approach described in Sect. 2 is data-driven and does not require the participation of experts in defining the weights of evaluation criteria. However, in IES and distance learning the professional expertize is needed in many aspects on – definition of needs and targets, selection of criteria, and application of proper assessment methods [18]. In this paper are not explore topics related to expert activities which are well discussed in different publications [19, 20].

The conventional AHP allows to express and structure the preferences of DMs and users about their perceptions for needs, design, performance, costs, etc. This is achieved by comparing criteria in pairs whose total number is defined as "$n(n-1)/2$". For example, for 3 criteria the total number of "pair comparisons" is $3(3-1)/2 = 3$, but the rising of criteria number increases exponentially the number of hierarchy comparisons. For this study, the systematization of the 9 independent criteria in 3 blocks decreases the number of comparisons for 1 expert 12 times (from 36 to 3) as shown in Table 4.

Table 4. Number of criteria/blocks and pairs of hierarchy comparisons in AHP

Number of criteria or blocks "n"	2	3	4	5	9	10	15	20
Comparisons for 1 DM: $n(n-1)/2$	1	3	6	10	36	45	105	190
Comparisons for 2 DMs: $2n(n-1)/2$	2	6	12	20	72	90	210	380
Comparisons for 4 DMs: $4n(n-1)/2$	4	12	24	40	144	180	410	760

Our studies are focused on improving the traditional "Analytical Hierarchy Process (AHP) method [21, 22] approaches by decomposing the initial set of criteria into smaller logical blocks for optimizing the assessment in different techniques. In the algorithm of Sect. 2, an additional "Step 1.0" is integrated for defining subjectively with the classical AHP method the weights of a limited number of blocks "b" ($b > 1$ and, in our opinion, preferably $b \leq 5$) with the. Such systematization organizes the expression of professional experts' opinions and frames the subjectivity of individual judgments up to the point of defining the importance of blocks. There-after, the definition of weight within the blocks depends on evaluation data. In our case, the initial criteria are grouped in 3 blocks: 1) "Learning (Block 1)" - for all teaching parameters (C_1, C_2, C_3, C_4); 2), "Practice (Bock 2)" - for parameters related to peer and self-quiz in practice (C_5, C_6, C_7); 3) "Exam (Block 3)"- for official final assessment (C_8, C_9).

The logical grouping of criteria reduces the number of pairs-comparisons and allows formulating consistently the preferences of DMs. Block weights are defined within a quadratic *Hierarchy Preference Matrix HPM$_{(b \times b)}$* which reflects the judgments of DMs according to the traditional AHP scale. The values of "1-3-5-7-9" quantify the string of basic qualitative labels: *"equal importance - moderate importance - strong importance - very strong importance - top importance"*. The interim levels (2-4-6-8) can be used for more precision. Such "AHP(blocks)" approach allows modelling the values in $HPM_{(b \times b)}$ with less computations and subjectivity. Block weights *(bw)* are calculated as:

$$bw_e = \left(\prod_{e=1}^{b} hp_{ee}\right)^{1/b} / \left(\sum_{e=1}^{b}\left(\prod_{e=1}^{b} hp_{ee}\right)^{1/b}\right), \text{ where } e = 1, \ldots, b. \quad (13)$$

$$\sum_{e=1}^{b} bw_e = 1 \quad (14)$$

For better results and simplification 3 consistent *HPM* are proposed in Table 5.

Table 5. AHP Hierarchy Preference Matrices (HPM) for block weights definition

Preferences	Exam				Learning & Exam				Learning & Practice			
Blocks	B1	B2	B3	Weight	B1	B2	B3	Weight	B1	B2	B3	Weight
B1:Learning	1	1	1/3	0.2	1	2	1	0.4	1	1/2	1/2	0.4
B2: Practice	1	1	1/3	0.2	1/2	1	1/2	0.2	2	1	1	0.4
B3: Exam	3	3	1	0.6	1	2	1	0.4	2	1	1	0.2
Consistency	CR = 0% (excellent)				CR = 0% (excellent)				CR = 0% (excellent)			

To define objectively the weights of criteria in the blocks with more than one criterion the Entropy transformation is calculated as described in Step 3 of Sect. 2 and finally the *sub-criteria weights "scw_j"* are defined by adjusting the entropy criteria weights (*recw_j* and *becw_j*) with the block weights bw_e:

$$rescw_j = bw_e.recw_j \quad (15)$$

$$descw_j = bw_e.decw_j \quad (16)$$

The *real and differed entropy weights* (*recw_j* and *becw_j*) as defined in Eq. (6–10) are unique for any set of criteria and alternative. In our case they are calculated in Table 3 and participate in the calculation of *sub-criteria weights* for all the blocks. The definition of weights for the preferences of "Exam", Learning & Exam", and "Learning & Practice" are presented respectively in Tables 6, 7 and 8.

In the hybrid "AHP-block & Entropy" approach all *"rescw"* are deducted directly from primary data to reflect the real distribution of information and adjusted with the block weights in each preference scenario.

In opposite, the artificial extraction for calculating *"descw"* provokes substantial redistribution of information resources between the blocks. In Block 1 the importance of "Theory Quiz" is reduced to neglecting levels: in the "Exam preference" from 0.052 to 0.003 (~17 times); in the "Learning & Exam preference" from 0.104 to 0,006 (~17

Table 6. Entropy sub-criteria weights for "Exam preferences" block

AHP Blocks	B1: Learning				B2: Practice			B3:	Exam
Block weights	bw = 0.2				bw = 0.2			bw = 0.6	
Criteria	A	A&P	TQ	TTQ	PA	SQ	TSQ	E	TE
e = SEnorm	0,869	0,986	0,998	0,999	0,989	0,994	0,997	0,957	0,999
recw	0,225	0,256	0,259	0,259	0,332	0,334	0,335	0,489	0,511
rescw	0,045	0,051	0,052	0,052	0,066	0,067	0,067	0,294	0,306
d = 1 - e	0,131	0,014	0,002	0,001	0,011	0,006	0,003	0,043	0,001
decw	0,889	0,092	0,015	0,004	0,570	0,298	0,132	0,984	0,016
descw	0,178	0,018	0,003	0,001	0,114	0,060	0,026	0,590	0,010

Table 7. Entropy sub-criteria weights for "Learning & Exam preferences" block

AHP Blocks	B1: Learning				B2: Practice			B3:	Exam
Block weights	bw = 0.4				bw = 0.2			bw = 0.4	
Criteria	A	A&P	TQ	TTQ	PA	SQ	TSQ	E	TE
e = SEnorm	0,869	0,986	0,998	0,999	0,989	0,994	0,997	0,957	0,999
recw	0,225	0,256	0,259	0,259	0,332	0,334	0,335	0,489	0,511
rescw	0,090	0,102	0,104	0,104	0,066	0,067	0,067	0,196	0,204
d = 1 - e	0,131	0,014	0,002	0,001	0,011	0,006	0,003	0,043	0,001
decw	0,889	0,092	0,015	0,004	0,570	0,298	0,132	0,984	0,016
descw	0,356	0,037	0,006	0,002	0,114	0,060	0,026	0,394	0,006

Table 8. Entropy sub-criteria weights for "Learning & Practice preferences" block

AHP Blocks	B1: Learning				B2: Practice			B3:	Exam
Block weights	bw = 0.4				bw = 0.4			bw = 0.2	
Criteria	A	A&P	TQ	TTQ	PA	SQ	TSQ	E	TE
e = SEnorm	0,869	0,986	0,998	0,999	0,989	0,994	0,997	0,957	0,999
recw	0,225	0,256	0,259	0,259	0,332	0,334	0,335	0,489	0,511
rescw	0,090	0,102	0,104	0,104	0,133	0,133	0,134	0,098	0,102
d = 1 - e	0,131	0,014	0,002	0,001	0,011	0,006	0,003	0,043	0,001
decw	0,889	0,092	0,015	0,004	0,570	0,298	0,132	0,984	0,016
descw	0,356	0,037	0,006	0,002	0,228	0,119	0,053	0,197	0,003

times); and in the "Practice & Exam preference" from 0.104 to 0,006 (~17 times). Redistributions, although in smaller scales, are observed in Block 2 and Block 3. Different kind of data treatments are possible in principle, but in real situations, the researchers should be concerned about their content and repercussions on the facts and conclusions.

4 Evaluation with TOPSIS, MOORA and WPM

Taking into account that the different evaluation techniques are based on different concepts of data treatment, logically, they cannot produce identical results on the same sets of empirical data. In this sense, to receive credible results more than one technique should be used in scientific research and real MDCM processes. The choice of evaluation techniques depends on the experience and preferences of researchers. In this study are applied 3 of the most popular techniques which contain different mathematical logics and can assess both "profit" and "cost" criteria – TOPSIS, MOOR, and WPN.

In parallel with the evaluations and rankings, this study performed an analysis based on the Spearman correlation statistics which is traditionally regarded as a reliable tool for testing the robustness and effectiveness of different indicators.

4.1 Technique for Order Preference by Similarity to Ideal Solution (TOPSIS)

TOPSIS is a popular approach with a compensatory concept for handling complex tasks for the assessment of numerous alternatives with multiple controversial criteria [23, 24]. It is based on the principle of the shortest Euclidean distances to the ideal positive solution and the largest distance from the ideal negative solution by identifying the "Value best rating" (v_j^+) and the "Value worst rating" (v_j^-) for all target alternatives in all criteria. For "profit" criteria $v_j^+ = \max(nwr_{ip})$, and $v_j^- = \min(nwr_{ip})$. For "cost" criteria (C_c), $v_j^+ = \min(nwr_{ic})$, and $v_j^- = \max(nwr_{ic})$.

Step 5. Calculation of TOPSIS Euclidian distances *(ed)* for each alternative as worst Sm_i^- and best Sm_i^+ separation measures in each criterion *"Cj"*:

$$Sm_i^- = \left(\sum\nolimits_{i=1}^{n} \left(v_{ij} - v_{ij}^- \right)^2 \right)^{-2} \tag{17}$$

$$Sm_i^+ = \left(\sum\nolimits_{i=1}^{n} \left(v_{ij}^+ - v_{ij} \right)^2 \right)^{-2} \tag{18}$$

Calculation of *"Similarity value"* of ideal worst solution *"Sv_i^*"*:

$$Sv_i^* = Sm_i^- / (Sm_i^+ + Sm_i^-) \tag{19}$$

The rankings of results with TOPSIS in the four variants of weighs for the 3 different selection preferences are displayed in Table 9.

In the variant of "AHP(blocks) & Entropy" with "recw" the results are identical for the variants "Exam" and "Learning & Exam" (A1 > ...A5 > A6 > A4 > A9 > A8 > A7) except for the 2nd and 3rd position. Logically, the selection is different in the case of "Learning & Practice" preference (A10 > A3 > A1 > A2 > A4 > A5 > A6 > A9 > A7 > A8).

4.2 Multi-objective Optimization Based on Ratio Analysis (MOORA)

MOORA is a popular and intuitive approach for multi-criteria optimization and evaluation [25, 26] thanks to its additive and easy for computation mathematical logic. At the same time, thanks to the compensatory concept it can handle complex tasks including multiple controversial criteria.

Table 9. Final ranking of TOPSIS assessment results for selection of students

TOPSIS	AHP(blocks) & Entropy						Entropy	
Weights	rescw			descw			recw	decw
Block preference	E	L&E	L&P	E	L&E	L&P	no preference	no preference
A1	1	1	3	1	1	1	2	1
A2	2	3	4	3	3	3	4	3
A3	4	4	2	4	4	4	3	4
A4	7	7	5	7	6	5	5	6
A5	5	5	6	5	7	9	7	9
A6	6	6	7	6	5	6	6	5
A7	10	10	9	10	9	8	9	8
A8	9	9	10	9	10	10	10	10
A9	8	8	8	8	8	7	8	7
A10	3	2	1	2	2	2	1	2

Legend: E – Exam preference; L&E – Learning & Exam preference; P&E– Practice & Exam preferences"

Step 5. Calculation of MOORA Alternatives Assessments (MAA_i) for each alternative.

$$MAA_i = \sum_{j=1}^{g} p_{ij} r_{ij} - \sum_{j=g+1}^{n} p_{ij} r_{ij} \qquad (20)$$

where "r_{ij}" is calculated as per Eq. (3).

The rankings of results with MOORA in the four variants of weighs for the 3 different preferences are displayed in Table 10.

Table 10. Final ranking of MOORA assessment results for selection of students

MOORA	AHP (blocks) & Entropy						Entropy	
Weights	rescw			descw			recw	decw
Block preference	E	L&E	L&P	E	L&E	L&P	no preference	no preference
A1	1	1	2	1	1	2	1	1
A2	3	3	4	3	3	4	4	3
A3	4	4	3	4	4	3	3	4
A4	6	6	5	7	5	5	5	5
A5	5	5	6	5	7	7	6	8
A6	7	8	8	6	6	6	8	6
A7	10	9	9	10	9	9	9	9
A8	9	10	10	9	10	10	10	10
A9	8	7	7	8	8	8	7	7
A10	2	2	1	2	2	1	2	2

Legend: E – Exam preference; L&E – Learning & Exam preference; P&E– Practice & Exam preferences

In the MOORA variant of *"AHP(blocks) & Entropy"* with *"recw"* the results are identical for "Exam" and "Learning & Exam" for the 6 best alternatives (A1 > A10 > A2 > A3 > A5 > A4) and very similar for the rest of the selections. The rankings with *"decw"* are identical for the 4 best and the 6th and 8th ranked alternatives. Logically, they differ from "Learning & Practice" (A10 > A3 > A1 > A2 > A4 > A5 > A6 > A9

> A7 > A8). In general, the selections in MOORA are more similar to the selections in the case of WPM than in in the case of TOPSIS.

4.3 Weighted Product Method (WPM)

WPM is also a very accessible method with multiplicative logic allowing to perform selections with "Benefit/Profit" and "Non-Benefit/Cost" criteria [27–29].

Step 5. Calculation of *Product Alternatives Assessments "PAA"* for alternatives A_i:

$$PAA_i = \prod_{j=1}^{n} a_{ij}^{cw_j} \tag{21}$$

– for "Benefit" criteria a_{ij} is defined as:

$$a_{ij} = x_{ij}/\max(x_{ij}) \tag{22}$$

– for "Non-Benefit" criteria a_{ij} is defined as:

$$a_{ij} = \min(x_{ij})/x_{ij} \tag{23}$$

The rankings of results with WPM in the four variants of weighs for the 3 different selection preferences are displayed in Table 11.

Table 11. Finals ranking of WPM assessments for students' performance

WPM	AHP (blocks) & Entropy						Entropy	
Weights	recw			decw			recw	decw
Block preference	E	L&E	L&P	E	L&E	L&P	no preference	no preference
A1	1	1	2	1	1	2	1	1
A2	3	3	4	3	3	3	3	3
A3	4	4	3	4	4	4	4	4
A4	6	6	5	7	6	5	5	5
A5	5	5	6	5	7	7	6	8
A6	7	8	8	6	5	6	8	6
A7	10	10	9	10	10	9	10	9
A8	9	9	10	9	9	10	9	10
A9	8	7	7	8	8	8	7	7
A10	2	2	1	2	2	1	2	2

Legend: E – Exam preference; L&E – Learning & Exam preference; P&E– Practice & Exam preferences

In the WPM variant of *"AHP(blocks) & Entropy"* with *"recw"* the results are identical for "Exam" and "Learning & Exam" for the 6 best alternatives (A1 > A10 > A2 > A3 > A5 > A6) and very similar for the rest of the selections. The rankings with *"decw"* are less similar – only for the best 4 alternatives (A1 > A10 > A2 > A3).

The selection in WPM with *"recw"* are identical with the results in MOORA and very similar to those in TOPSIS. A detailed comparative analysis of all evaluation techniques on the basis of correlation is presented in the following Sect. 5.

5 Correlation Analysis of TOPSIS, MOORA and WPM Results

One of the main directions in our studies is the exploration of the effects of real entropy end "differed entropy" on the weights of criteria and on the assessment results. The stand-alone entropy method is totally data-driven and does not allow to model the preferences between the criteria. In this case the difference between "recw" and "decw" is determined by the primary dataset. Table 12 contains the correlation tests about the rankings in each of 3 evaluation techniques - TOPSIS, MOORA, and WPM.

The correlation in the variant of stand-alone Entropy is different but always lower than in each evaluation technique indicating that the evaluation values and rankings become more perturbed when *"decw"* are applied. When AHP(blocks) systematization is applied the correlations in the "Exam" preference are at the highest levels. The lowest correlation in TOPSIS is received in the "Learning & Practice" preference, while the lowest correlations for MOORA and WPM are obtained for the "Learning & Exam" preference, except for the case of "Practice preference". A logical reason for this kind of difference, in our opinion, can be searched in the different mathematical logics and the fact that cumulative concept in MOORA and the multiplicative in WPM are closer to each other and differ from the "Euclidean concept" in TOPSIS. This is as an indication that TOPSIS is expected to extend the difference between "real entropy" and "differed entropy". However, all techniques contain formal mathematical rules and the results of evaluation depend on the sets with real primary data and not on the area of application.

Another important indicator about the effectiveness of the AHP(blocks) approach is the correlation of results between the different preference cases. The more similar block weights in the pairs of "Exam" v/s "Learning & Exam" and in the "Learning & Exam" v/s "Learning & Practice" produce more similar ranking which are confirmed by the higher correlations values than in the pair of "Exam" v/s "Learning & Practice". This can be regarded as a confirmation that the Euclidian logic in TOPSIS is expected to produce more differentiated results which can lead to less similarity with the assessments and ranking in the other methods.

In general, the primary datasets of evalution parameters for the education process are usually correlated and can be used for predicting with different models the final performance of students [30, 31]. In our case, the correlation values in the variant of "real weights" are higher which is an indication about the consistency of the model and the simulation on the dataset.

Another important observation is that the hybrid "AHP(blocks) & Entropy" approach based on "real entropy weights" produces more clear and similar results in the three evolution methods. This contributes to design and support more robust MCDM processes for obtaining more credible results. Due to their more similar logics MOORA and WPM produce more similar results in all three variants of preferences. In this particular case, the similarity of TOPSIS with MOORA and WPM is insignificantly lower, with a slightly closer resemblance between TOPSIS and WPM. Such comparisons are very important for cross-validating the results in each of the technique and for their eventual integration in a single final assessment.

The correlations between the different evaluation techniques are also an indicator about their behavior in the variants of real and differed entropy and the positive role

Table 12. Spearman correlation of rankings in TOPSIS, MOORA and WPN

Correlation: rescw v/s descw	AHP (blocks) & -SE (with preference)		SE (no preference)	
TOPSIS "E"	0,988		0,927	
TOPSIS "L-E"	0,952			
TOPSIS "L-P"	0,867			
MOORA "E"	0,988		0,939	
MOORA "L-E"	0,939			
MOORA "L-P"	0,964			
WPM "E"	0,983		0,917	
WPM "L-E"	0,883			
WPM "L-P"	0,942			
Correlations of preferences	AHP&SE (recw)	AHP&SE (decw)	SE (recw)	SE (decw)
TOPSIS: E v/s L-E	0,988	0,952	1	
TOPSIS: E v/s L-P	0,855	0,842		
TOPSIS: L-E v/s L-P	0,891	0,952		
MOORA: E v/s L-E	0,976	0,939	1	
MOORA: E v/s L-P	0,940	0,915		
MOORA: L-E v/s L-P	0,964	0,976		
WPM: E v/s L-E	0,983	0,950	1	
WPM: E v/s L-P	0,933	0,918		
WPM: L-E v/s L-P	0,949	0,964		

Legend: E - Exam preference; L-E – Learning & Exam preference; L-P – Learning & Practice preference.

of the combined "AHP(blocks) & Entropy" method for producing more consolidated results that reflect the preferences of users as displayed in Table 13.

One of the purposes of our studies is to explore the repercussions of the "differed entropy" generated in Eq. (9–10) on the robustness of the MCDM process. The advantages of the combined "AHP(blocks) & Entropy" method are confirmed by simulating selections in three different types of preferences. The systematization of criteria in blocks sets clear limits for the redistribution of information caused by "differed entropy" within the defined "quotas" for each block, while an artificially definition of weights disregards the real education processes and risks to corrupt the MCDM evaluation at its final stage.

A major advantage of including peer and self-assessment aspects in the IES is the possibility to motivate the students to involve more actively and effectively in all stages of the education process for improving theoretical knowledge, practice skills, and team responsibility which is particularly important under conditions of distance learning and social containment. The dataset employed in the simulations of this study includes alternatives

Table 13. Correlations of rankings in different preference cases

Correlations of rankings in "Exam preference"				
Correlations (Spearman)	recw (AHP&SE)	recw (SE)	decw(SE)	decw (AHP&SE)
TOPSIS v/s MOORA	0,976	0,952	0,976	1
TOPSIS v/s WPM	0,976	0,927	0,976	1
MOORA v/s WPM	1	0,976	1	1
Correlation of rankings in "Learning & Exam preference"				
TOPSIS v/s MOORA	0,952	0,952	0,976	0,988
TOPSIS v/s WPM	0,964	0,927	0,976	0,988
MOORA v/s WPM	0,988	0,976	1	0,976
Correlations of rankings in "Learning & Practice preference"				
TOPSIS v/s MOORA	0,976	0,952	0,976	0,939
TOPSIS v/s WPM	0,976	0,927	0,976	0,952
MOORA v/s WPM	1	0,976	1	0,988

that were ranked without major controversies in the variants with different preferences. A real situation with more alternatives and criteria is expected to make the evaluation more complicated and would require further development of Artificial Intelligence aspects in the IES.

6   Back-End AHP for Multiple Assessment Integration (MAI)

When several evaluation techniques are performed it is useful to consolidate their results for mitigating eventual differences and facilitating the final choice of users and DMs [32]. A possible approach is to calculate a weighted sum of assessments results in two ways – on the basis of rankings R_i or on the basis of assessment values AV_i. We can call this approach a "Multiple Assessment Integration (MAI)".

$$MAI(R_i)\sum_{q=1}^{q} w_i R_i \tag{24}$$

$$MAI(AV_i) = \sum_{q=1}^{q} w_i AV_i \tag{25}$$

where "q" is the number of evaluation techniques used in the MCDM process, and "w" is the weight of each technique in the different variant of MAI.

Here again, the AHP method can be used to define the weights of the different evaluation techniques in a more correct and transparent way. For demonstration purposes this paper presents an approach with 3 different variants of "*MAI*" shown in Table 14.

On one side, the integration of rankings "*MAI(R_i)*" seems to be more simple and intuitive since all results are in the same evaluation interval – in our case this interval ($R_i \in 1 - 10$) is defined by the number of alternatives "$m = 10$". On this basis, the

Table 14. AHP definition of weights for evaluation techniques in MAI

Preferences	MAI A				MAI B				MAI C			
Blocks	T	M	W	Weight	T	M	W	Weight	T	M	W	Weight
TOPSIS	1	1	1	0.33.	1	8	8	0.8	1	47	47	0.96
MOORA	1	1	1	0.33.	1/8	1	11	0.1	1/47	1	1	0.02
WPM	1	1	1	0.33.	1/8	2	1	0.1	1/47	1	1	0.02
Consistency	CR = 0% (excellent)				CR = 0% (excellent)				CR = 0% (excellent)			

importance of different evaluation techniques can be defined transparently by the relative weights and does not depend on the content of the assessment's values (AV_i). As this approach is very simple and intuitive it is not discussed in this section. On other side, the integration of assessment values "$MAI(AV_i)$" is a more challenging task since the different evaluation techniques produce sets of assessment values (AV_i) which in most of the case are situated in different intervals. In this situation, to model a transparent and predictable "$MAI(AV_i)$" the researcher should analyze the numerical content of assessment values. For demonstration purposes in this paper the integration is presented for the variant of "Exam preference".

In "$MAI\ A$" all evaluation techniques are allocated with equal importance which results in an even distribution of relative weights, and logically, the final integrated ranking is identical with the stand-alone variant of WPM and MOORA. The domination of WPM can be explored with sensitivity analysis by gradually increasing the importance of other evaluation techniques, as show in Table 15. For demonstration purposes, we perform this analysis on the basis of TOPSIS, as the results in MOORA are practically identical with WPM. In "$MAI\ B$" we increase formally 8 times the weight importance of TOPSIS but still obtain the same final integrated assessment rankings as in "$MAI\ A$" with the domination of WPM. The domination of WPM continues up to the point when the hierarchy preference of TOPSIS in the AHP matrix is increased to 47 times in respect to WPM and MOORA as shown in "$MAI\ C$".

It should be noticed, that the evaluation of students' performance in the education process have several important specifics. For instance, the datasets in different countries may contain results that are based on evaluation systems with very different number of grades (i.e. from 3 or 4 to 100). This will produce very different results after normalization - datasets based on smaller number of grades will remain less differentiated which will embarrass further ranking and classifications. This type of "equalization effect increases with the number of alternatives (in our case - students) included in the surveys. For that reason, in such evaluation systems the increasing of the number of criteria will have little effect on diversification for larger population even if AHP(blocks) or other methods are applied. In opposite, in less equalitarian evalution systems the multi-criteria methods can combine more successfully the professional expertise with the objectivity of facts.

In this sense, "MAI" can be a convenient tool in the cases when more controversial datasets will have to compared in the MCDM process. Still, it has to be take into

Table 15. Multiple Assessment Integration (MAI): Exam preference

Method	TOPSIS		MOORA		WPM		MAI A		MAI B		MAI C	
Alternative	AV_i	R_i	AV_i	R_i	AV_i	R_i	AV_i	R_i	AV_i	R_i	AV_i	R_i
A1	**0,924**	**1**	**0,081**	**1**	**0,949**	**1**	**0,651**	**1**	**0,875**	**1**	**0,907**	**1**
A2	0,890	2	0,064	3	0,847	3	0,601	3	0,839	3	0,873	2
A3	0,808	4	0,053	4	0,818	4	0,560	4	0,765	4	0,793	4
A4	0,514	7	0,002	6	0,680	6	0,398	7	0,493	7	0,506	7
A5	0,727	5	0,030	5	0,755	5	0,504	5	0,688	5	0,713	5
A6	0,613	6	-0,008	7	0,662	7	0,422	6	0,580	6	0,602	6
A7	0,128	10	-0,075	10	0,388	10	0,147	10	0,131	10	0,129	10
A8	0,328	9	-0,061	9	0,544	9	0,270	9	0,318	9	0,324	9
A9	0,466	8	-0,017	8	0,636	8	0,361	8	0,447	8	0,459	8
A10	0,889	3	0,070	2	0,918	2	0,625	2	0,842	2	0,873	3
Max	*0,924*		*0,081*		*0,949*		*0,65*		*0,842*		*0,907*	
Min	*0,128*		*-0,075*		*0,388*		*0,15*		*0,133*		*0,129*	
Max-Min	*0,796*		*0,156*		*0,561*		*0,50*		*0,708*		*0,778*	
Arith.Mean	*0,629*		*0,014*		*0,720*		*0,45*		*0,576*		*0,618*	

Legend: AV – Assessment Value; R – Rank.

account, that students are human beings with very different personal physical, intellectual and emotional characteristics, and, therefore, generalizations over different types of populations or period of time should be approached with care and attention.

7   Conclusions and Further Research

The methodological improvements and simulations in this study confirm that the enlargement of the stand-alone information entropy with the "block weighting" of criteria in AHP provides a logical and accessible approach for multi-criteria analysis as an effective support in the decision-making process. The comparative analysis of the variants with real and differed weights of criteria confirms the advantage for direct measuring of diversity and creates a basis for further research in other areas.

This study can be used as an initial vision for further development of multi-criteria assessment of students' performance in different IES. Peer and self-assessments can be useful supplements to the traditional methods of current and final examinations, which should continue to be the key road-stones in the assessment of education results. In this paper the number of students was reduced to 10 taking into account the of the limited volume of the publication format. Taking into account the accuracy of mathematical models the size of population is not limited from theoretical point of view and larger populations can be assessed in future applied studies with a reasonable definition of comparable samples and groups of students.

Future studies should enlarge the application of "AHP(blocks) & Entropy" approach with other evaluation methods, such as VIKOR, ELECTREE, etc. The Multiple Assessment Integration (MAI) can be used in different areas as a convenient tool for consolidating the final results from different assessments.

The objective approach for criteria weighting can be improved with the introduction of novel and reliable indicators for concentration and hierarchy as a logical enlargement and supplements to the traditional entropy.

Acknowledgements. This research is supported by the Bulgarian FNI fund through the project "Modeling and Research of Intelligent Educational Systems and Sensor Networks (ISOSeM)", contract КП-06-Н47/4 from 26.11.2020.

References

1. Wiener, N.: Cybernetics: or Control and Communication in the Animal and the Machine, 2nd revised edn. Paris, (Hermann & Cie) & Camb. Mass. MIT Press (1961). ISBN: 978-0-262-73009-9
2. Heims, S.J., Von Neumann, J., Wiener, N.: From Mathematics to the Technologies of Life and Death. MIT Press, Cambridge (1980)
3. Cumming, G., Self, J.: Intelligent educational systems: identifying and decoupling the conversational levels. Instrum. Sci. **19**, 11–27 (1990). https://doi.org/10.1007/BF00377983
4. Soller, A., Lesgold, A.: A computational approach to analysing online knowledge sharing interaction. In: Hoppe, U., Verdejo, F., Kay, J. (eds.) AI-ED 2003, pp. 253–260. IOS Press, Amsterdam (2003)
5. Lee, E.A.: The past, present and future of cyber-physical systems: a focus on models. Sensors **15**(3), 4837–5486 (2015)
6. Terzieva, V., Pavlov, Y., Todorova, K., Kademova-Katzarova, P.: Utility and optimal usage of ICT in schools. In: Rachev, B., Smrikarov, A. (eds.) ACM International Conference Proceeding Series, vol. 1369, pp. 302–309 (2017)
7. Boud, D., Falchikov, N. (eds.): Rethinking Assessment in Higher Education: Learning for the Longer Term. Routledge, London (2007)
8. Elander, J.: Student assessment from a psychological perspective. Psychol. Learn. Teach. **3**(2), 114–121 (2004)
9. Sivan, A.: Implementing peer assessment to enhance teaching and learning. New Horiz. **1**(2), 10–11 (2002)
10. Saaty, T.: Principia Mathematica Decernendi: Mathematical Principles of Decision Making. RWS Publications, Pittsburgh (2010)
11. Zavadskas, E., Turskis, Z.: Multiple criteria decision making (MCDM) methods in economics: an overview. Technol. Econ. Dev. Econ. **17**(2), 397–427 (2011)
12. Atanassova, V., Doukovska, L., Karastoyanov, D., Čapkovič, F.: Inter-criteria decision making approach to EU member states competitiveness analysis: trend analysis. In: Proceedings of the 7th IEEE International Conference on Intelligent Systems – IS 2014, vol. 1, Warsaw, Poland, 24–26 September 2014. Mathematical Foundations, Theory, Analyses. Springer International Publishing, Switzerland, AISC, vol. 322, pp. 107–115 (2016)
13. Kirilov, L., Guliashki, V., et al.: An overview of multiple objective job shop scheduling techniques. JÖKULL J. **66**(2), 172–206 (2016)
14. Ilchev, S., Andreev, R., et al.: Ultra-compact laser diode driver for the control of positioning laser units in industrial machinery. In: Stapleton, L., et al. (ed.) IFAC Papers Online, vol. 52, issue number 25, pp. 435–440 (2019)
15. Shannon, C.: A mathematical theory of communication. Bell Syst. Tech. J. **27**(3), 379–423 (1948)
16. Petrov, I.I.: Block criteria systematization with AHP and entropy-MOORA approach for MCDM in selecting desktop PCs. In: Proceedings of Conference "TechSys 2021", 8 p. (2021, in print). http://techsys.tu-plovdiv.bg/index.html#HOME
17. Papinczak, T., Young, L., Groves, M., Haynes, M.: An analysis of peer, self, and tutor assessment in problem-based learning tutorials. Med Teach **29**(5), 122–132 (2007)

18. Cho, K., MacArthur, C.: Student revision with peer and expert reviewing. Learn. Instr. **20**(4), 328–338 (2010)
19. Saaty, T., Peniwati, K.: Group Decision Making: Drawing Out and Reconciling Differences. RWS Publications, Pittsburgh (2008)
20. Borissova, D.: A group decision making model considering experts' competency: an application in personnel selections. Comptes rendus de l'Academie Bulgare des Sci. **71**(11), 1520–1527 (2018)
21. Saaty, T.: Decision Making for Leaders: The Analytic Hierarchy Process for Decisions in a Complex World. RWS Publications, Pittsburgh (2008)
22. Forman, E., Saul, G.: The analytical hierarchy process—an exposition. Oper. Res. **49**(4), 469–487 (2001)
23. Hwang, C.-L., Yoon, K.: Multiple Attribute Decision Making: Methods and Applications. Springer-Verlag, New York (1981). https://doi.org/10.1007/978-3-642-48318-9
24. Wang, T.C., Lee, H.D.: Developing a fuzzy TOPSIS approach based on subjective weights and objective weights. Expert Syst. Appl. **36**(5), 8980–8985 (2009)
25. Brauers, T., Zavadskas, E.: Robustness of the multi-objective MOORA method with a test for the facilities sector. Technol. Econ. Devel. Econ. **15**(2), 352–375 (2009)
26. Ghorabaee, K.M., Zavadskas, E.K., Olfat, L., Turskis, Z.: Multi-criteria inventory classification using a new method of evaluation based on distance from average solution (EDAS). Informatica **26**(3), 435–451 (2015)
27. Triantaphyllou, E., Mann, S.H.: An examination of the effectiveness of multi-dimensional decision-making methods: A decision-making paradox. Decis. Support Syst. **5**(3), 303–312 (1989)
28. Borissova, D., Keremedchiev, D.: Assessing and ranking students by multi-attribute decision making model based on SMART. Cybern. Inf. Technol. **19**(3), 45–56 (2019)
29. Şahin, M.: A comprehensive analysis of weighting and multi-criteria methods in the context of sustainable energy. Int. J. Environ. Sci. Technol. **18**, 1591–1616 (2021)
30. Dicheva, D., Dichev, C., Agre, G., Angelova, G.: Gamification in education: a systematic mapping study. Educ. Technol. Soc. **18**(3), 75–88 (2015). ISSN: 1176-3647
31. Gray, C.C., Perkins, D.: Utilizing early engagement and machine learning to predict student outcomes. Comput. Educ. **131**, 22–32 (2019)
32. Hussain, A., Chun, J., Khan, M.: A novel customer-centric methodology for optimal service selection (MOSS) in a cloud environment. Futur. Gener. Comput. Syst. **105**, 562–580 (2020)

Modeling the Association Between Prenatal Exposure to Mercury and Neurodevelopment of Children

Stefan Popov[1,2](✉) [iD], Janja Snoj Tratnik[3] [iD], Martin Breskvar[2] [iD],
Darja Mazej[3], Milena Horvat[1,3] [iD], and Sašo Džeroski[1,2] [iD]

[1] Jožef Stefan International Postgraduate School, Jamova cesta 39,
Ljubljana, Slovenia
{stefan.popov,milena.horvat,saso.dzeroski}@ijs.si
[2] Department of Knowledge Technologies, Jožef Stefan Institute, Jamova cesta 39,
Ljubljana, Slovenia
martin.breskvar@ijs.si
[3] Department of Environmental Sciences, Jožef Stefan Institute, Jamova cesta 39,
Ljubljana, Slovenia
{janja.tratnik,darja.mazej}@ijs.si

Abstract. This work presents an application of machine learning methods in the area of environmental epidemiology. We have used lifestyle and exposure data from 769 mother-child pairs from Slovenia and Croatia to predict the neurodevelopment of the children, expressed through five Bayley-III test scores. We have applied single- and multi-target (semi-)supervised predictive methods to build models capable of predicting the Bayley-III scores. Additionally, we have used feature ranking methods to estimate the importance of individual lifestyle and mercury exposure attributes on the Bayley-III test scores. The learned models offer useful insights into the effect of prenatal mercury exposure on the neural development of children.

Keywords: Machine learning · Multi-target regression ·
Environmental epidemiology · Feature ranking

1 Introduction

Mercury (Hg) is known to have adverse impacts on human health [5]. The general population is mainly exposed to mercury in two ways: (1) through the diet - mostly by fish consumption (methyl Hg) and (2) through dental amalgam fillings (Hg° vapour). Prenatal or early postnatal exposure to methyl Hg can cause neurodevelopmental disorders in children. A recent study [11] investigates the association between prenatal exposure to mercury and neurodevelopment of children, taking into account gene data (apolipoprotein E-*Apoe*). For their purpose they have surveyed mother-child pairs from the central region in Slovenia and from Rijeka, a city on the Croatian coast in the northern Adriatic, and have collected data on their lifestyle and Hg exposure. The neurodevelopment of some

© Springer Nature Switzerland AG 2022
L. Antovski and G. Armenski (Eds.): ICT Innovations 2021, CCIS 1521, pp. 85–97, 2022.
https://doi.org/10.1007/978-3-031-04206-5_7

children at 18 months of age has been assessed with the Bayley Scales of Infant and Toddler Development, Third Edition (Bayley-III) Test [2]. This test helps to identify children with delay in development and assesses their development in different domains.

This study focuses on the data set provided by the PHIME study [11]. Our goal is to train machine learning models that will be able to predict the Bayley-III scores from the lifestyle and exposure data.

2   Data

The PHIME project [11], is part of a larger longitudinal birth cohort study set in the Mediterranean area. The project was designed to investigate the association between prenatal mercury exposure from fish consumption during pregnancy and neuropsychological development of children as well as to investigate the co-exposure to other potentially neurotoxic elements and their role in biological response of the children exposed to Hg in the prenatal period.

The PHIME project started with the recruitment of women in their last trimester of pregnancy or at child birth. The collected data consists of 540 mother-child pairs from Slovenia, and 229 from Croatia. At birth, cord blood and maternal scalp hair were sampled for determination of trace elements concentrations. Mothers filled out a brief inclusion questionnaire, including general information about health, dietary habits and socio-economic status. Six to eight weeks later, breast milk was collected by mothers. Mothers were also required to fill out a detailed questionnaire regarding their health, life-style and dietary habits, socio-economic status, residential and occupational history. The children were followed up at 18 months of age for assessment of their neuropsychological performance using Bayley Scales of Infant and Toddler Development, Third Edition Test (Bayley, 2006), administering cognitive, language and motor (fine and gross) scores:

1. Cognition composite score (CCS)
2. Language composite score (LCS)
3. Motor composite score (MCS)
4. Fine Motor scaled score (FMSS)
5. Gross Motor scaled score (GMSS)

At the time of testing, another (supplementary) questionnaire was filled out by the mothers, including the type of feeding from birth onwards, and behavioural features. Table 1, taken from the workbook on Bayley-III scores, summarizes the information about their values.

The data set is rather incomplete as there are a lot of missing data. For example, the blood and urine features are available only for the Croatian population, which makes up for less than 30% of our data set. For 331 mother-child pairs there are no Bayley-III scores available. In total, the data set consists of 769 mother-child pairs which are described with 82 descriptive attributes (lifestyle and exposure data). Our goal is to predict the values of 5 target attributes (Bayley scores).

Table 1. Descriptive classification of the Bayley target scores

Composite score or equivalent	Class
130 and above	Very superior
120–129	Superior
110–119	High average
90–109	Average
80–89	Low average
70–79	Borderline
69 and below	Extremely low

3 Machine Learning Methods

The availability, dimensionality and type of the target variables that we are trying to predict (Bayley-III scores) determine the machine learning task. Given that there are multiple numerical (integer) target variables, and the fact that not all instances (mother-child pairs) have known values for them, the task at hand is semi-supervised multi-target regression (MTR).

Generally, MTR problems can be approached in two ways: locally or globally. When a local approach is used, one predictive model is learned for each target attribute. Alternatively, when using a global approach to MTR, a single model is learned that is able to predict values for all targets simultaneously. The difference between the two approaches is in the way how the target space is interpreted by the algorithm. When using the local approach, no potential relations between the target attributes can be exploited as the algorithm only focuses on one target attribute. With global approaches, the potential relations between the target attributes are taken into account and can, in some cases, lead to better predictive performance. In cases, when interpretable model types are used, such as decision trees and decision rules, the global approach yields a single interpretable model, as opposed to several interpretable models resulting from the local approach to MTR. It can be challenging for the domain experts to combine local models into an overall interpretation. In this work, we apply both local and global approaches and compare their predictive performance.

Simple models (models with low complexity in terms of how they interpret the input space) often exhibit low predictive power. It is a standard practice to combine many such models into ensembles, which is a known way of improving the predictive performance. An ensemble model combines predictions of the individual models within the ensemble to produce the final prediction. Such ensemble models are also often used for feature ranking.

A feature ranking is a list of all descriptive attributes (inputs), ordered according to their ranking scores. The idea is to determine which descriptive attributes carry the most discriminating information w.r.t. the target attribute(s), i.e., the higher the ranking score of a given descriptive attribute, the higher its importance. In settings with high-dimensional descriptive spaces, it is often beneficial to reduce the number of descriptive attributes before learn-ing the predictive models. Obviously, removing highly important attributes will result in poor predictive performance. Hence, by using a feature ranking algo-rithm, one can determine the importance of all attributes and make an educated cut-off (ultimately considering only the attributes with high importance scores). The ranking of attributes can also be used to validate interpretable predictive models, i.e., if an attribute appears high in a decision tree and also has a high feature ranking score, one can be confident that that attribute is quite important w.r.t. discriminating the values of the target attribute(s).

In this study, we have used machine learning methods that produce inter-pretable predictive models as well as ensembles thereof. Additionally, we have also used a feature ranking method to produce a ranking of descriptive attributes. The used methods are briefly described below. All methods are implemented within the CLUS[1] software. The first type of interpretable models that we used are predictive clustering trees (PCTs). In particular, we have built MTR trees [12], such as the ones shown in Figs. 1 and 2. PCTs are based on the predictive clustering paradigm [3], which generalizes decision trees and parametrizes them to support multitude of structured output prediction tasks, one of which is MTR. Decision trees can also be seen as a hierarchical clustering, where the structure of the decision tree mirrors the clustering hierarchy. Each node represents a cluster that can be described by the tests that appear in the tree. Each node holds a test and if we combine all the tests from the root node to the selected node, we get the description of the cluster at the selected node. A prediction with a PCT is made in the same way as with a standard decision tree.

The importance scores were calculated by using the feature ranking method for MTR [8]. This method is based on ensembles of MTR trees [6] and calculates the Genie3 importance score, based on Random forests (RFs) of 100 PCTs for MTR. The importance scores and the corresponding ranking denote the relative importance of each attribute for predicting all targets, jointly and separately. Highly ranked attributes contain the most discriminative information w.r.t. the target(s) of choice.

Our data set contains missing values for many of the target attributes, i.e., not all mother-child pairs have known values for the Bayley-III scores. The standard PCT top-down induction algorithm does not support such cases. Therefore, we have also used semi-supervised PCTs (SSL-PCTs), an extension to the standard PCT induction algorithm, where both, labeled and unlabeled instances are used for calculating the heuristic score of candidate splits during model learning [7].

Both, PCTs and SSL-PCT were also used in the ensemble setting. In par-ticular, we have used the RF algorithm to build our ensembles of (SSL-)PCTs. The RF algorithm builds an ensemble of many decision trees in order to lift the predictive performance over that of individual PCTs in the ensemble. The RF ensembles with PCTs and SSL-PCTs are denoted as RF-PCTs and RF-SSL-PCTs, respectively.

---

[1] CLUS software is available for download at http://source.ijs.si/ktclus/clus-public.

The second type of interpretable models we used are predictive clustering rules (PCRs) [16]. PCRs are multi-target decision rules, capable of modeling MTR problems. The PCR algorithm implements the standard sequential covering algorithm for rule discovery. In each step, the standard covering algorithm generates a single rule and removes data instances from the data set which are covered by that rule. A data instance is *covered* by a rule if it satisfies its condition clause. The algorithm continues to generate rules until there are no more instances left in our data set. A rule is added to the rule set if the predictive performance of the rule set with the new rule is better than without it. When making predictions, the discovered rules can be used in one of two ways: ordered or unordered. When rules are ordered (such models are often called decision lists), only one rule can be triggered. The order of the rules is determined by the algorithm. If none of the rules are triggered, the default rule is applied. The triggered rule gives the final prediction. This explicitly gives higher importance to those rules that have a higher weight, which can affect the interpretation of the predictions. When using unordered rules, several rules can be triggered, i.e., the instance, for which the predictions are being produced, can, depending on the rule conditions, trigger more than one rule. In those cases, predictions are combined into the final prediction (similar to what is done with tree ensembles).

4   Related Work

The predictive clustering framework has been successfully applied to many diverse problems in the domain of life and medical sciences. Here we name a few. [15] have applied predictive clustering methods to reveal the relationship between fungi and different salt concentrations. Their study has revealed new interesting properties about halophilic fungi and has expanded the knowledge of possible life performance under diverse and extreme environmental conditions. [4] have utilized the clustering aspect of PCTs and have discovered interesting clusters of patients with Alzheimer's disease that share biological features. The clusters have discovered both gender specific differences and several biological features that can relate to the progression of the disease. [14] have used PCTs to identify subgroups of patients with Parkinson's disease that would react positively or negatively to medication modification. Their findings will assist physicians that make the therapy modifications for a given patient by narrowing down the number of possible scenarios.

In the recent works by [1,11,13] multiple linear regression has been applied to evaluate possible relationship between Hg exposure in prenatal life and 5 neurodevelopmental scores of children at 18 months of age. The model adjusted for potential confounders (mother's age, child's sex, birth weight, education of the mother, smoking during pregnancy and concentration of selenium and lead in cord blood) revealed that doubling the Hg concentration on cord blood would result in 0.33 points lower fine motor score. Similar decrement was observed for Slovenian and Croatian populations in the meta-analysis done by [1]. On the other hand, doubling the Hg concentration in cord blood of Apoe ϵ4 carriers would decrease the cognitive score for 5.4 points [11]. The observed changes were small on an individual level, but were statistically significant and relevant on a global (population) scale.

To the best of our knowledge, there is no publication related to the application of machine learning methods to the problem of associating prenatal and early postnatal exposome with the neural development of children. Given the geographic specificity of the problem, and its potential to generalise to the entire human population, we consider this publication to be very relevant in the field of environmental epidemiology.

5   Experimental Setup

PCT and SSL-PCT models were built with a variance reduction heuristic and M5Multi pruning method [10]. Same setup was used for both single- and multi-target variants. In the standard PCT top-down induction (TDI) algorithm, the variance reduction heuristic is calculated based on the values of the target variables. Our data set contains some instances where values of the target variables are not known. Therefore, the standard PCT TDI algorithm needs to be instantiated with a different variance reduction heuristic function. In particular, SSL-PCTs introduce the w parameter, used to control the contribution of target and descriptive attributes variances towards the overall variance in the currently observed instances. This parameter is data set sensitive and must be optimized [7] for each data set individually. Therefore, we optimize it by using 5-fold internal cross-validation to select one of the candidate values which range between 0.1 and 1.0 with a step of 0.1.

To build the random forest ensemble models (for prediction and feature ranking) we used 100 individual (SSL-)PCTs as base learners. Each (SSL-)PCT was allowed to grow without limiting the number of instances in the leaf nodes, i.e., no pre-pruning was applied, and had only a subset of $sqrt|D|$ random attributes available when learning, where $|D|$ is the number of descriptive attributes. The final prediction of the ensemble is obtained by taking the predictions of the individual (SSL-)PCTs and calculating their arithmetic mean.

Ordered PCRs were learned by using the standard covering algorithm, adding additional rules only if they improve the predictive performance of the model. Unordered PCRs were learned by using the weighted covering algorithm, where the only difference from the former algorithm is that we do not immediately remove instances that are covered by a new rule, but rather decrease their weight inversely proportional to the error that the new rule makes when predicting their target values. For both rule-based models we used multiplicative dispersion search heuristic and added rules to the resulting rule set if and only if they cover at least 45 instances. Unordered PCRs were obtained by setting the weight controlling the amount by which weights of covered instances are reduced within the error weighted covering algorithm, to 0.5 and the instance's weight threshold to 0.1 (if an instance's weight falls below this value, it is removed from the learning set).

We calculated root relative squared errors (RRSEs) to evaluate the predictive performance of the generated models. RRSE is relative to what it would have been if we had just predicted the average value for each score. Thus, the relative squared error takes the total squared error from our model and normalizes it by dividing it with the total squared error of a model that simply predicts the average. In general, we want the RRSEs to be lower than one and as close to zero as possible. The formula for calculating RRSE for the target attribute t is:

$$\text{RRSE}_t = \sqrt{\frac{\sum_{i=1}^{N}(y_i - \hat{y}_i)^2}{\sum_{i=1}^{N}(y_i - \overline{y}_i)^2}}, \tag{1}$$

where N is the number of data points, y_i is the true target value of instance i, $\hat{y}_i$ is the predicted value for the target and $\overline{y}_i$ is the arithmetic mean, calculated over the target values within the training set. The average RRSE over T target attributes is then calculated as:

$$\text{aRRSE} = \frac{1}{T}\sum_{t=1}^{T}\text{RRSE}_t. \tag{2}$$

We used 10-fold cross-validation to estimate the RRSEs of our models. Table 4 contains the obtained RRSEs values.

6   Results

The algorithms for building PCTs and SSL-PCTs yield models that can easily be interpreted. The produced PCT and SSL-PCT models are shown in Fig. 1 and Fig. 2, respectively. The PCT model identified the child's gender, the concentration of methyl Hg in the mother's blood and the mother's age as the most relevant attributes. The semi-supervised PCT model identified the concentration of methyl Hg in the cord blood and the number of pregnancies as most relevant attributes.

The PCRs model illustrated in Table 2 consists of an ordered list of 9 rules. Each data instance that we are trying to predict is tested against the condition clause in the rules in the specified order. Prediction is done by the first rule that has its condition clause satisfied (i.e. the rules are ordered). If there exists no such rule, than the prediction from the default rule is applied.

Similarly, in Table 3 we illustrate an unordered PCRs model. There, the collection of rules can be seen as a set rather than a list. An instance is tested against each rule and a prediction is obtained by averaging the predictions from all individual rules that had their condition satisfied by the instance. If the instance fails to satisfy any condition, then the prediction from the default rule is taken as final.

Table 4 summarizes the values of root relative squared errors (RRSE) for each method per target score.

The random forest of PCTs with Genie3 feature ranking method outputs a list of attributes ordered by their importance scores. Each score is calculated

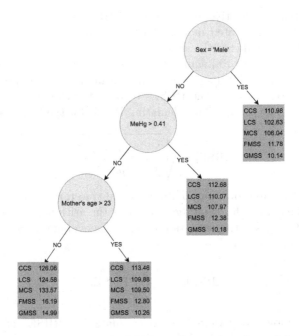

Fig. 1. A predictive clustering tree for MTR predicting the values of the five Bayley-III scores simultaneously.

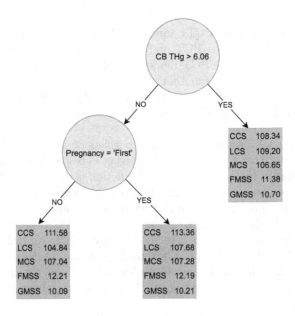

Fig. 2. A semi-supervised predictive clustering tree for MTR predicting the values of the five Bayley-III scores simultaneously.

Table 2. A list of PCRs for MTR predicting the five Bayley-III scores simultaneously. The rule conditions are given in the second column. The predictions are in columns CCS, LCS, MCS, FMSS and GMSS.

#	Rule conditions	CCS	LCS	MCS	FMSS	GMSS
1	CB_Zn ≤ 1507.795	117.06	106.73	107.50	12.32	10.04
2	CB_Serum_Ca > 2.89 AND CB_Serum_Mg > 0.71	113.26	106.15	108.69	12.69	10.08
3	M_milk_Mn ≤ 1.599	111.19	106.89	106.89	12.10	10.06
4	CB_As ≤ 0.466	114.34	105.32	106.76	12.36	9.78
5	GA > 40 AND CB_Se ≤ 112.38	112.17	105.34	106.13	12.17	9.78
6	M_milk_Zn ≤ 2539.085	108.04	104.83	106.56	12.00	10.13
7	no_amalgams > 2 AND M_hair_THg > 26	114.02	107.43	107.02	12.21	10.06
8	frozenfish = 3 AND gest_age = 2 AND BMI > 18.9069	106.41	105.39	107.26	11.60	10.71
9	CB_Serum_Ca > 2.51 AND BMI ≤ 33.91	115.65	112.34	110.84	12.32	11.23
10	Default	103.33	101.91	102.79	10.87	10.00

Table 3. A set of PCRs for MTR predicting the five Bayley-III scores simultaneously. The rule conditions are given in the second column. The predictions are in columns CCS, LCS, MCS, FMSS and GMSS.

#	Rule conditions	CCS	LCS	MCS	FMSS	GMSS
1	CB_Zn ≤ 1507.795	117.06	106.73	107.50	12.32	10.04
2	CB_Serum_Ca > 2.91 AND CB_Serum_FeIII ≤ 44.10	114.12	107.12	107.74	12.65	9.81
3	M_milk_Mn ≤ 1.397 AND M_milk_Se > 8.344747	110.27	106.48	106.87	12.14	10.04
4	Default	111.70	106.38	107.19	12.09	10.21

as the average decrease in impurity while inducting the PCTs. The greater the decrease, the more important an attribute is. Table 5 gives a snapshot of the list, showing only the top and bottom three attributes as well as some other attributes in between that appear in the (SSL-)PCT models.

The ranking in Table 5 identifies the concentration of total Hg in the mother's hair as the most important attribute for determining the Bayley-III scores, the concentration of Cu (copper) and As (arsenic) in the cord blood as the second and third most important attribute, and continues down to the information about whether the mother is smoking currently, the concentration of methyl Hg

Table 4. The RRSEs for each multi-target model the an overall average RRSE (aRRSE) across all five scores. Bolded numbers denote the best-performing model for a given score. Underlined number denotes the best overall-performing model.

Method	Cognition	Language	Motor	Fine motor	Gross motor	Overall
PCT	1.0094	0.9961	1.0158	1.0071	1.0147	1.0086
SSL-PCT	1.0094	1.0130	**1.0096**	1.0070	1.0016	1.0081
Ordered PCRs	**0.9836**	1.0168	1.0117	1.0095	1.0136	1.0070
Unordered PCRs	0.9942	1.0048	1.0019	1.0029	**0.9944**	0.99968
RF-PCTs	0.9855	**0.9928**	1.0118	**0.9888**	1.0098	<u>0.9977</u>
RF-SSL-PCTs	0.9868	1.0072	1.0113	0.9931	1.0051	1.0007

Table 5. Feature ranking of the top ten/bottom three attributes and of those that appear in the (SSL-)PCT models.

Rank	Attribute name	Importance score
1	M_hair_THg	806.3
2	CB_Cu	786.8
3	CB_As	769.3
4	mothers_age	768.4
5	CB_Pb	762.5
6	CB_MeHg	697.5
7	freshfish	680.6
8	BMI	672.6
9	CB_Se	665.4
10	CB_THg	663.9
11	child_sex	605.3
41	M_blood_MeHg	123
54	firstpregnancy	90.8
...	...	...
80	smokingcurrently	9.7
81	M_milk_MeHg	5.9
82	numberofcigarettesperday	1.9

in her milk and the number of cigarettes she smokes per day as the three least important factors. Mother's age, child's sex and the concentration of total Hg in the cord blood can also be considered very important, because they appear in the tree models, and rank very high in the rankings.

The above described results are in line with the main outcomes of the Slovenian and Croatian birth cohort study PHIME, which tested the existence of association between exposure to methyl Hg in prenatal or early life and neurodevelopmental performance of children. There have been reports [9,11,13] on significantly negative association between total or methyl Hg in cord blood or maternal hair (both indicate exposure to Hg in prenatal period) and fine motor scores of Bayley-III assessment, as well as cognitive scores, although only in a sub-population of carriers of apolipoprotein epsilon 4 gene variant [11,13]. Prenatal exposure to Hg was confirmed as a significant predictor for cognitive and fine motor scores regardless of the genotype by the RF-PCT+Genie3 and SSL-PCT methods, ranking at the first position in the former method, and being the root in the tree model in the latter. However, some additional predictors were revealed in the present study, namely copper (Cu) and arsenic (As) concentrations in cord blood, the first known for its pro-inflammatory effects and the second for potential neurotoxicity, similarly as Hg. Both Hg and As share the source of exposure which might explain the observed significance. First pregnancy also came out as an important attribute, which is yet to be explained (Table 6).

Table 6. The RRSEs for each single-target model and the overall average RRSE (aRRSE) across all five scores. Bolded numbers denote the best-performing model for a given score. Underlined number denotes the best overall-performing model.

Method	Cognition	Language	Motor	Fine motor	Gross motor	Overall
PCT	1.0157	**0.9745**	1.0067	1.0226	1.0212	1.0081
SSL-PCT	1.0221	1.0092	1.0040	1.0051	**1.0000**	1.0080
Ordered PCRs	1.0147	1.0364	1.0066	1.0134	1.0099	1.0162
Unordered PCRs	0.9993	1.0113	**1.0004**	1.0083	1.0013	1.0041
RF-PCTs	0.9921	0.9895	1.0165	0.9934	1.0158	1.0014
RF-SSL-PCTs	**0.9909**	0.9892	1.0099	**0.9896**	1.0112	<u>0.9981</u>

In our particular case, local and global approaches exhibit similar predictive performance. Local approach outperforms the global only for the best language and motor score, and, for other scores, including the overall one, the global approach is marginally better. SSL-PCTs perform slightly better than fully supervised PCTs.

7 Conclusion

In this paper, we have applied machine learning methods to model the associations between exposure to mercury in the environment and neural development of children. In this multi-target regression problem our target attributes represent Bayley-III scores. The problem was modeled with PCTs, RF of PCTs and semi-supervised variants of them, as well as with PCRs. We have also produced a ranked list of attributes, according to their importance when used for predicting the target attributes.

All methods generate models with comparable predictive performance but the best performing model was generated with the RF-PCTs method. Given the specific nature of the problem, an observation can be made that a global approach is better than a local one, because it generalizes better and thus captures only the high-level relationships between the features, and does not succumb to the noise introduced by the missing data and limited number of instances. The random forest of PCTs model marginally outperformed the baseline model (simply predicting the average value) for the targets Cognitive score, Language score and Fine motor (scaled). Similarly, the random forest of SSL-PCTs outperformed the baseline model for the targets Cognitive score and Fine motor (scaled). PCRs and PCTs were able to predict one target (Language and Cognition score, respectively) better than the simple baseline model. Other models were not able to outperform the baseline model, predicting arithmetic mean for individual targets, calculated on the training data. We believe that the poor predictive performance of generated models can be attributed to high sparsity of the data set. Obtaining more labeled data should result in better performing models.

Despite this rather low predictive power of models, the results obtained are in line with the main findings of previous work on the PHIME data set. Some additional predictors were revealed, providing valuable insight into the environmental epidemiology aspects of chronic low-level exposures and will be further evaluated by using an expanded version of the data set. Application of machine learning methods is particularly valuable in studies like PHIME, where a large number of attributes is used to make a prediction within a rather narrow range of values.

Acknowledgements. SP would like to acknowledge the financial support in the form of a scholarship of the Public Scholarship, Development, Disability and Maintenance Fund of the Republic of Slovenia. SD, MB and SP would like to acknowledge the grant number P2-0103 (the research programme Knowledge Technologies). All authors acknowledge the NEURODYS project (J7-9400, Neuropsychological dysfunctions caused by low level exposure to selected environmental pollutants in susceptible population) for financial support of the overall work. JST, DM and MH also acknowledge the EU funded 6th FP project PHIME for providing the data used in this work.

References

1. Barbone, F., et al.: Prenatal mercury exposure and child neurodevelopment outcomes at 18 months: results from the Mediterranean Phime cohort. Int. J. Hyg. Environ. Health **222**(1), 9–21 (2019). https://doi.org/10.1016/j.ijheh.2018.07.011
2. Logsdon, A.: Bayley Scales of Infant and Toddler Development. 3rd edn (2008). http://images.pearsonclinical.com/images/pdf/bayley-iii_webinar.pdf, last accessed 07.09.2020
3. Blockeel, H., De Raedt, L., Ramon, J.: Top-down induction of clustering trees. In: Proceedings of the Fifteenth International Conference on Machine Learning, pp. 55–63 (1998)
4. Breskvar, M., Zenko, B., Džeroski, S.: Relating biological and clinical features of Alzheimer's patients with predictive clustering trees. In: International Multi-Conference Information Society (2015)

5. Kim, K.H., Kabir, E., Jahan, S.A.: A review on the distribution of hg in the environment and its human health impacts. J. Hazard. Mater. **306**, 376–385 (2016). https://doi.org/10.1016/j.jhazmat.2015.11.031
6. Kocev, D., Vens, C., Struyf, J., Džeroski, S.: Tree ensembles for predicting structured outputs. Pattern Recogn. **46**(3), 817–833 (2013)
7. Levatić, J., Kocev, D., Ceci, M., Džeroski, S.: Semi-supervised trees for multi-target regression. Inf. Sci. **450**, 109–127 (2018). https://doi.org/10.1016/j.ins.2018.03.033
8. Petković, M., Kocev, D., Džeroski, S.: Feature ranking for multi-target regression. Mach. Learn. **109**(6), 1179–1204 (2019). https://doi.org/10.1007/s10994-019-05829-8
9. Prpic, I., et al.: Prenatal exposure to low-level methyl mercury alters the child's fine motor skills at the age of 18 months. Environ. Res. **152** (2016). https://doi.org/10.1016/j.envres.2016.10.011
10. Quinlan, J.R.: Learning with continuous classes. In: 5th Australian Joint Conference on Artificial Intelligence, pp. 343–348. World Scientific (1992)
11. Snoj Tratnik, J., et al.: Prenatal mercury exposure, neurodevelopment and apolipoprotein e genetic polymorphism. Environ. Res. **152**, 375–385 (2017). https://doi.org/10.1016/j.envres.2016.08.035
12. Struyf, J., Džeroski, S.: Constraint based induction of multi-objective regression trees. In: International Workshop on Knowledge Discovery in Inductive Databases, pp. 222–233. Springer, Berlin (2005). https://doi.org/10.1007/978-3-540-75549-4
13. Trdin, A., et al.: Mercury speciation in prenatal exposure in Slovenian and Croatian population - Phime study. Environ. Res. **177**, 108627 (2019). https://doi.org/10.1016/j.envres.2019.108627, https://www.sciencedirect.com/science/article/pii/S0013935119304244
14. Valmarska, A., Miljkovic, D., Konitsiotis, S., Gatsios, D., Lavrac, N., Robnik-Sikonja, M.: Combining multitask learning and short time series analysis in Parkinson's disease patients stratification. In: Conference on Artificial Medicine in Europe, pp. 116–125, May 2017. https://doi.org/10.1007/978-3-319-59758-4_13
15. Zajc, J., et al.: Chaophilic or chaotolerant fungi: a new category of extremophiles? Front. Microbiol. **5** (2014)
16. Ženko, B.: Learning predictive clustering rules. Ph.D. thesis, University of Ljubljana (2007), http://eprints.fri.uni-lj.si/709/1/zenko%2Dphd%2Dthesis.pdf

DD-RDL: Drug-Disease Relation Discovery and Labeling

Jovana Dobreva, Milos Jovanovik$^{(\boxtimes)}$, and Dimitar Trajanov

Faculty of Computer Science and Engineering, Ss. Cyril and Methodius University
in Skopje, Skopje, North Macedonia
{jovana.dobreva,milos.jovanovik,dimitar.trajanov}@finki.ukim.mk

Abstract. Drug repurposing, which is concerned with the study of the effectiveness of existing drugs on new diseases, has been growing in importance in the last few years. One of the core methodologies for drug repurposing is text-mining, where novel biological entity relationships are extracted from existing biomedical literature and publications, whose number skyrocketed in the last couple of years. This paper proposes an NLP approach for drug-disease relation discovery and labeling (DD-RDL), which employs a series of steps to analyze a corpus of abstracts of scientific biomedical research papers. The proposed ML pipeline restructures the free text from a set of words into drug-disease pairs using state-of-the-art text mining methodologies and natural language processing tools. The model's output is a set of extracted triplets in the form (drug, verb, disease), where each triple describes a relationship between a drug and a disease detected in the corpus. We evaluate the model based on a gold standard dataset for drug-disease relationships, and we demonstrate that it is possible to achieve similar results without requiring a large amount of annotated biological data or predefined semantic rules. Additionally, as an experimental case, we analyze the research papers published as part of the COVID-19 Open Research Dataset (CORD-19) to extract and identify relations between drugs and diseases related to the ongoing pandemic.

Keywords: Drug-disease relations · NLP · Knowledge extraction

1 Introduction

Drug discovery is a time-consuming, expensive, and high-risk process and it usually takes 10–15 years to develop a new drug, with the success rate of developing a new molecular entity being only around 2% [31]. Because of that, studying drugs that are already approved to treat one disease to see if they are effective for treating other diseases, known as drug repurposing, has been growing in importance in the last few years [27]. The core methodologies of drug repurposing approaches can be divided into three categories: network-based approaches, text-mining approaches, and semantic approaches [31]. The text-mining approach's main goal is the processing of medical and biological literature and extracting

L. Antovski and G. Armenski (Eds.): ICT Innovations 2021, CCIS 1521, pp. 98–112, 2022.
https://doi.org/10.1007/978-3-031-04206-5_8

fruitful novel biological entity relationships. The importance of the text-mining approach has been increasing in the last couple of years due to the fact that the amount of scientific publications related to the biomedical and life sciences domain is growing at an exponential rate, with more than 32 million medical publications available on PubMed. On the other hand, the latest advancements in natural language processing (NLP) have significantly improved the efficiency of language modeling [4]. Text mining in the medical and biological domain can be divided into four phases [31]: information retrieval (IR), biological name entity recognition (BNER), biological information extraction (BIE), and biological knowledge discovery (BKD). In the IR step, the relevant documents are selected from the identified source. In the BNER step, the biological entities are identified in each of the retrieved documents. This step is accomplished with the use of the NLP subtask for named entity recognition (NER) [16]. NER includes the automated implementation of natural language processing tasks like tokenization, handling punctuation, stop terms and lemmatization, as well as the ability to create custom (case-specific) text processing functions like joining consecutive tokens to tag a multi-token object or performing text similarity computations. In other words, we can tell that NER uses the grammatical sentence structure to highlight the given entities.

The task of dealing with the syntactic rules in the sentences is a part of co-reference resolution [24]. Co-reference resolution, i.e. the replacement of all expressions referring to the same entity in the text, is an important step for many higher-level NLP tasks which involve understanding the natural language, such as summarizing documents, answering questions, and extracting information [25]. On the other hand, applying semantic role labeling (SRL) on the processed texts leads to observing the semantic part of the sentences and extracts the represented roles from it [29]. The BIE and BKD steps include the tasks of knowledge extraction from the biological corpus, such as drug-disease pairs and drug-action-disease triples. Based on our previous research experience in the field of applying NLP techniques to the biomedical domain [3,10], we recognized a need for a new model that will be able to learn new information about potential therapies for given diseases, based on published medical research. In this paper, we present a literature-based discovery (LBD) model which employs a series of steps to analyze the corpus of abstracts of scientific biomedical research papers. For that purpose, the model restructures the free text from a set of words into drug-disease pairs using the text mining techniques and the NLP tools. With this, it is able to build a network of known diseases and drugs which are correlated. The graph representation used by the model represents an abbreviated form of the knowledge that covers all drugs and diseases which are covered throughout the entire corpus of abstracts. In it, each node represents a disease or a drug entity that is interconnected with other nodes via a weighted edge, which denotes the verb that occurs most frequently with the given pair. The inclusion of verbs in the process allows us to add an additional label to the relation, which renders the knowledge extraction more precise. We used datasets which are publicly available and represent publications from the medical domain. Therefore, the corpus was very large, but the data was not labeled or annotated. In order to use the raw data and get comparable results we used NLP state-of-the-art models, such as AllenNLP, BERT, RoBERTa, GTP, etc. The evaluation of our model is based on the gold standard dataset for drug-disease cases that contains 360 drug-disease relationships which are selected from the therapeutic target database [22]. As an experimental case, we also analyze the COVID-19 by analyzed the abstracts of the research papers published as part of the COVID-19 Open Research Dataset (CORD-19) [26].

2   Related Work

Applying text mining and NLP methods to discover drug-disease pairs has been an exciting research field in recent years. In [27] the authors propose a system that extracts the drug-disease pairs by using the vocabulary of drugs and diseases from the Unified Medical Language System (UMLS). Due to the large drug and disease dictionary size, an optimized variant of the string searching algorithm Aho-Corasick [2] is used. The algorithm is a dictionary-matching algorithm that locates elements of a finite set of strings (dictionary) within an input text and matches all strings simultaneously. The learned patterns are then used to extract additional treatment and inducement pairs. A similar approach is used in [30], where the extracted patterns are detected from the sentences that have the following format "DRUG Pattern DISEASE" or "DISEASE Pattern DRUG". For example, the top five patterns are: "in", "in the treatment of", "for", "in patients with", "on", and we can note that some of them do not express the real nature of the relationship. Based on literature mining, the creators of [12] built a knowledge base that is used to find biological entities linked to the COVID-19 disease. They collect disease-drug interactions from the CORD-19 literature and categorize them as positive or negative (labels). The positive label indicates that the medicine is moderately effective in curing the condition, whereas the negative label indicates that the drug is ineffective in curing the ailment. The proposed platform does not find more details regarding the relationship between drugs and diseases and how they are connected. The authors of [17] expand the pattern-based technique beyond single drugs and diseases to drugs combinations. They look for trends in the relationship between drug combinations and diseases. They use a pattern-based technique to extract illness and medication combination pairings from MEDLINE abstracts and build a word meaning disambiguation system based on POS tagging. Using many language characteristics, including lexical and dependency information, the authors of [8] proposed a supervised learning technique for automatically extracting chemical-induced disease connections. Furthermore, the suggested technique makes use of the MeSH restricted vocabulary to aid in the training of classifiers and to solve the issue of relation redundancy during the extraction process. Using gold-standard entity annotations, the system obtained F-scores of 58.3% on the test dataset.

The authors of [19] investigate the many sorts of relationships that exist in LBD systems. They utilize the basic A-B-C model [9], where six types of relations are covered in this overview: c-doc, c-sent, c-title, SemRep, ReVerb, and the

Stanford parser. C-doc looks for term co-occurrence over the whole document (in this case, a document is an abstract). C-sent takes a more stringent approach, considering phrases to be co-occurring if they appear in the same sentence inside a document (abstract) and c-title just examines document titles. SemRep [13] is a free accessible tool that uses underspecified syntactic processing and UMLS domain knowledge to extract subject-relation-object triples (such as "X treats Y") from biomedical literature. Based on enforced syntactic and lexical constraints, the freely accessible ReVerb information extraction technology extracts binary relations conveyed by verbs. The Stanford parser extracts phrase structure trees and creates typed grammatical connections, such as subject, between pairs of words. For the purpose of assessment, the authors create a gold standard dataset. With time slicing, this is possible: hidden knowledge is created from all data up to a given cut-off date and compared against fresh ideas provided in publications after the cut-off date. Identifying innovative ideas in publications after the cut-off date, on the other hand, is not straightforward: extracting all newly co-occurring pairs of CUIs, for example, will obviously provide a huge and noisy gold standard, favoring LBD number over quality.

SemRep [13] is an NLP system that uses linguistic principles and UMLS domain knowledge to extract semantic relations from PubMed abstracts. The assessment is based on two different datasets. They employ a manually annotated test collection and undertake thorough error analysis in one study[1]. SemRep's performance on the CDR dataset, a typical benchmark corpus annotated with causal chemical-disease correlations, is also evaluated. On a manually annotated dataset, a rigorous assessment of SemRep provides 0.55 precision, 0.34 recall, and 0.42 F_1 score. A more lenient evaluation provides 0.69 accuracy, 0.42 recall, and 0.52 F_1 score, which more properly represents SemRep performance.

SemaTyP [22] is a biomedical knowledge graph-based drug discovery approach that mines published biomedical literature to find potential medicines for illnesses. They first use SemRep to create a biomedical knowledge graph from the relationships extracted from biomedical abstracts, then train a logistic regression model by learning the semantic types of paths of known drug therapies that exist in the biomedical knowledge graph, and finally use the learned model to discover drug therapies for new diseases. They provide a gold standard[2] of drug-disease instances for the assessment from TTD[3]. They chose TTD's 360 drug-disease correlations as the gold standard for drug rediscovery testing. NEDD [33] is a computational technique based on meta-paths. Using Hin2Vec [5] to generate the embeddings, they first create a heterogeneous network as an undirected graph by combining drug-drug similarity, disease-disease similarity, and known drug-disease correlations. The low dimensional representation vectors of medications and diseases are generated by NEDD using meta pathways of various lengths to explicitly reflect the indirect links, or high order closeness, inside drugs and diseases. Experiments using a gold standard dataset [7], which contains 1,933

---

[1] https://semrep.nlm.nih.gov/GoldStandard.html.

[2] https://doi.org/10.6084/m9.figshare.6389870.v1.

[3] http://db.idrblab.net/ttd/.

verified drug-disease correlations, demonstrate that NEDD outperforms state-of-the-art methods in terms of prediction outcomes. The authors offer a technique in [21] that differentiates seven connections between two semantic concepts, "therapy" and "disease". Five graphical models and a neural network are represented. Only three relations were found, namely, "cure", "prevent", and "side effect", which were represented with accuracy levels of 92.6, 38.5, and 20, respectively. The papers [1,11,21,32] represent different classification models for relationships between drugs and diseases. In each of them the main idea is to first apply a named entity recognition for the medical terms and then to use approaches based on linguistic patterns and domain knowledge. For the classification part, the authors choose diverse types of ML models, such as neural networks, SVM, logistic regression, etc.

The proposed methods for drug-disease pairs discovery are mainly lexicon-based and do not enable discovering the actions (the verbs) connecting the disease and the drug. Starting from this drawbacks we propose a model that, besides discovering the drug-disease pairs, also allows for labeling the relationship with the verb that connects the detected entities. The models that are used in the proposed ML pipeline are based on publicly available state-of-the-art NLP models that are mainly trained on standard English text. With the obtained results, we have shown that we can get comparable results without using large amounts of annotated biomedical data or specially coded grammatical and semantic rules.

3   Drug-Disease Relation Discovery and Labeling (DD-RDL)

The main goal is to build a text mining tool for biomedical abstracts which will provide useful knowledge extraction with the help of general NLP techniques. Our model, unlike other models, does not use CUIs or UMLS, and is solely based on entity extraction and building a co-occurrence matrices. Therefore, our experiment is based on trying various NLP tools and techniques which will lead to a good LBD system, so the focus is on exploring and testing if the newest text mining tools will prove useful in the biomedical field. On the other hand, building and training a whole new domain-specific model especially when we are building a graph, takes a lot of time. Therefore, our model uses domain specific and globally used NLP tools for faster knowledge extraction, where a big data corpus is not necessary for achieving good accuracy. The state-of-the-art models for this problem are building NLP tools that are domain specific. Our idea is to try already existing NLP models for English language into the medical field, we know that the words and meanings are different into the spoken English and medically based one. In other word, the training of each NLP task from the DD-RDL model is on English language, but we are transferring that knowledge into domain specific one in order to see is the relation extraction just as successful as in the language tasks. The Drug-Disease Relation Discovery and Labeling (DD-RDL) model is composed of six stages grouped in two subsystems (Fig. 1). The first subsystem is composed of a pipeline that has three steps, starting with named entities recognition (NER), continuing with co-reference resolution (CR), and ending with semantic role labeling (SRL). The second subsystem is composed of knowledge extraction tasks, and its pipeline is comprised of three stages: verb grouping, adjacency matrix construction, and knowledge extraction. The code is publicly available on GitLab[4].

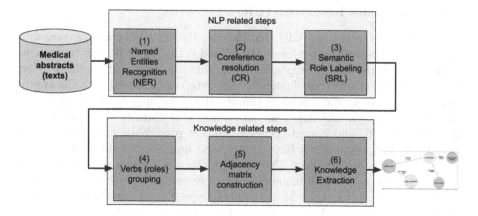

Fig. 1. Structure of the DD-RDL Model.

As an input, the DD-RDL model gets a dataset of medical or life science-related abstracts of papers, or full texts. As a result, the model provides a set with triplets in the form (drug, verb, disease) with its relative frequency as an additional information. The model also generates a knowledge graph where nodes are disease and drugs, connected with links that represent the connecting verbs (roles) and the associated relative frequency. The entire process is available through a module that can be installed and started from a command-line interface, but we also provide a web service where the client can attach multiple scientific papers and process them with the DD-RDL model.

3.1 Named Entity Recognition (NER)

The first step is finding entities from the medical domain, such as disease, syndrome, pharmaceutical substance, amino acid, protein, therapy, etc. For this purpose, we use the MedCAT library [14] that is designed to be used for extracting information from electronic health records (EHRs) and link them to biomedical ontologies, such as UMLS. MedCAT offers a pre-trained model that automatically builds a pallet of various medical entities, such as pharmaceutical substances, diseases, treatments, amino acid, peptides, proteins, etc. The first step is very important, because the globally used NER tools extract a different pallet of words, since these domain-specific terms do not occur in their bag-of-words.

---

[4] https://gitlab.com/jovana.dobreva16/dd-rdl-model.git.

Therefore, we used MedCAT which is able to highlight the biomedical domain entities and extract them based on their category. In order to improve the drug[5] and disease[6] detection, we add an additional step that uses a simple lookup-based algorithm, tables are available on gitlab[7], that is trying to find additional drugs that were not identified by the MedCAT library.

3.2 Co-reference Resolution (CR)

The co-reference resolution (CR) step includes finding all linguistic expressions (mentions) in the text that refer to the related entity. After finding those mentions, they are resolved by replacing them with the associated entities. After this step, the pronouns ("that", "she", "his", etc.) are replaced with the appropriate entity making the sentences simpler and more appropriate for the next step of semantic role labeling. There are a couple of libraries like StanfordNLP [20], AllenNLP [6], and Neural-Coref [15, 28] that provide an implementation of co-reference resolution algorithms. Due to the simplicity to use and the high accuracy, we use Neural-Coref. Even though these co-reference models are not the best possible solution for biomedical texts, given that they show lower performance results than the ones which are domain-specific, we want to have a model which is able to extract knowledge with the help of grammatical and semantic knowledge from general texts.

3.3 Semantic Role Labeling (SRL)

The goal of SRL is to identify the events in the sentence, such as discovering "who" did "what" to "whom", "where", "when", and "how". The predicates (typically verbs) are the central part of the SRL process, and they define "what" took place. The other sentence constituents express the participants in the event (such as "who" and "where"), as well as other event properties (such as "when" and "how"). SRL's major goal is to properly represent the semantic relationships that exist between a predicate and its related participants and attributes. These connections are chosen from a set of probable semantic roles for that predicate that has been pre-defined (or class of predicates) [18]. The SRL process is done by using a BERT-based algorithm for semantic role labeling [23]. When the input text passes through this step, the subject, verb, and object for each sentence is recorded. The subject and the object are then filtered to keep only those connected with the identified entities in the previous step. As a result of this step, we get a set of triples (subject, verb, object) associated with each input text.

---

[5] https://www.kaggle.com/arpikr/uci-drug.

[6] https://www.kaggle.com/priya1207/diseases-dataset.

[7] https://gitlab.com/jovana.dobreva16/dd-rdl-model/-/tree/master/data_storage.

3.4   Verbs Grouping

After processing the subject and object entities, the next step in the pipeline is verbs grouping. This step aims to lower the number of different verbs by replacing similar verbs with a single one that has the same meaning. First, all the verbs are converted to present tense, and then they are encoded using word2vec vectors from the NLTK library. A similarity matrix between all verbs is created, and the most similar verbs are grouped together. We use cosine similarity with a threshold to create the groups. When the groups are made, the most frequent verb is chosen as a group representative and is used to replace all other occurrences of the other verbs from the group. This step gives us a smaller range of verb-actions between the diseases and drugs, leading to decreased bias and variance.

3.5   Adjacency Matrix Construction

After the triplets are created, we use them to construct the adjacency matrix that represents the probability of the occurrence of the pair (verb, disease) with a given drug:

$$P(Verb, Disease|Drug)$$

We then prune the matrix in order to decrease the number of zero values in it. The triples that remain are the nodes and edges of the knowledge graph. The weight of each edge is a calculated relative frequency of the corresponding triple.

In order to generate the adjacency matrix, we first created two matrices where, the first one represents the probability of a given disease occurring with a known verb-drug pair:

$$P(Disease|Verb, Drug)$$

and the second one represents the probability of a given verb occurring together with a given drug:

$$P(Verb|Drug)$$

By inverting the first matrix and then multiplying it with the second matrix, we arrive to the above mentioned final matrix.

3.6   Knowledge Extraction

As a final step in the processing pipeline execution, from the adjacency matrix, we constructed a bipartite knowledge graph in which the nodes are drugs and diseases, and the links between them represent the identified action (verb). To each link, a relative frequency is also associated. An example of a knowledge graph constructed by the DD-RDL model is shown in Fig. 2.

As we showed in the previous steps, we added a dose of domain-specific knowledge into models that are globally used for processing English texts. We combined domain specific NER and SRL together with co-reference resolution based on the English grammar. In the final steps, we used a statistical and mathematical knowledge for building the matrices.

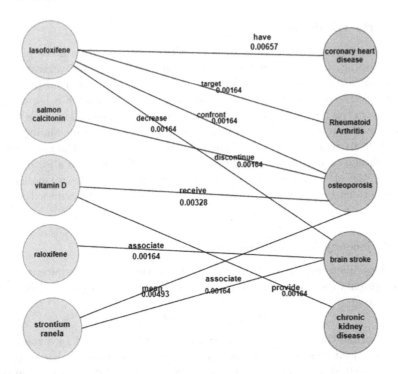

Fig. 2. An example knowledge graph of the knowledge extracted by the DD-RDL model.

4   Datasets

In order to compare our approach and its performance with existing systems, we used one of the gold standard datasets for drug-disease cases [22]. The dataset contains a pallet of different diseases out of which we selected two: osteoporosis and cardiovascular disease. The main reason of selecting just two diseases out of the corpus was because we wanted to test our model on a specific disease corpus, and see if the gold standard pairs were occurring in the model output, possibly along with new pairs. The gold standard dataset contains pairs of diseases and drugs that correlate based on previous medical research. Therefore, we needed to create a dataset that will represent a collection of abstracts where the given disease and drug occur. We collected a set of 203 abstracts from PubMed, which are related to the two selected diseases and the appropriate drugs paired with them from the drug-disease dataset. The selected papers refer to 27 different drug-disease pairs. The gold standard dataset [22] does not contain verb-actions for the drug-disease pairs, so we calculate the model's accuracy based only on the drug-disease coverage. In order to further test our model, we applied it on data from the COVID-19 Open Research Dataset (CORD-19) [26]. It represents a collection of scientific papers on COVID-19 and related historical coronavirus research. Even though it is composed of several features, we are simply interested in the title and the abstract of the papers. As we know, therapies for the SARS-CoV-2 virus are currently not defined. Although the result obtained using this dataset cannot be explicitly tested, we performed a manual check of the results.

5   Results

As we explained above, we use a set of 203 research paper abstracts from PubMed which contain 27 drug-disease pairs selected from the gold standard dataset for drug-disease cases [22]. The diseases in these pairs are osteoporosis and cardiovascular disease.

To show how the model extracts knowledge from the abstract, we will take a look at two sentences from two different abstracts, which are related to aspirin and cardiovascular disease. The sentences have the following representation:

- S1: "In 2016, the US Preventive Services Task Force recommended initiating aspirin for the primary prevention of both cardiovascular disease and colorectal cancer among adults ages 50 to 59 who are at increased risk for cardiovascular disease."
- S2: "Background: A polypill comprising statins, multiple blood-pressure-lowering drugs, and aspirin has been proposed to reduce the risk of cardiovascular disease."

From these two sentences the following SRL outputs were generated:

- S1: 'subject': 'the US Preventive Services Task Force', 'verb': 'recommend', 'object': 'initiating aspirin for the primary prevention of both cardiovascular disease and colorectal cancer among adults ages 50 to 59 who are at increased risk for cardiovascular disease'
- S2: 'subject': 'A polypill comprising statins, multiple blood-pressure-lowering drugs, and aspirin', 'verb': 'reduce', 'object': 'the risk of cardiovascular disease'.

The discovered (drug, verb, disease) triplet for the first sentence is (aspirin, recommend, cardiovascular disease) and for the second one is (aspirin, reduce, cardiovascular disease).

After running the abstracts through the pipeline of our model, the model extracted all pairs correctly, giving 100% coverage. Additionally, it discovered 18 new disease-drug pairs, which are correctly correlated. Given that the test dataset does not contain the action (verb), we calculated the accuracy based on the drug-disease pairs only, not the full triples. Some of the predicted pairs are shown in the Table 1, where we include the action verb to describe the drug-disease relation, and we show the calculated relative frequency. The relative frequency is calculated as the number of occurrences of the specified (drug, action, disease) triplet in the entire analyzed text corpus, divided by the total number of unique triplets extracted from the same corpus.

We performed the evaluation by comparing the (drug, disease) pairs we generated with the ones present in the gold standard dataset. We measured recall,

Table 1. A set of drugs, their actions on diseases and the relative frequency, extracted from the PubMed abstracts. We analyzed 203 paper abstracts from PubMed based on existing 27 drug-disease pairs, and managed to detect all of them with our model. This table presents 15 of them.

Drug	Action	Disease	Relative frequency
Calcium	Receive	Osteoporosis	0.00493
Vitamin E	Decrease	Cardiovascular Disease	0.00328
Strontium Ranelate	Present	Osteoporosis	0.00328
Ipriflavone	Prevent	Osteoporosis	0.00328
PTH-CBD	Prevent	Osteoporosis	0.00164
rhPTH	Evaluate	Osteoporosis	0.00164
Aspirin	Outweigh	Cardiovascular Disease	0.00164
Digitoxin	Block	Cardiovascular Disease	0.00164
Drug Combination	Underscore	Cardiovascular Disease	0.00164
Raloxifene	Decrease	Cardiovascular Disease	0.00164
Delapril	Ameliorate	Cardiovascular Disease	0.00164
Acarbose	Delay	Cardiovascular Disease	0.00164
Salmon Calcitonin	Discontinue	Osteoporosis	0.00164
Raloxifene	Prevent	Osteoporosis	0.00164
Lasofoxifene	Prevent	Osteoporosis	0.00164

which showed 100% accuracy. Other metrics, such as precision and F_1 score, cannot be utilized in our case, because we discover novel relations between drugs and diseases, relations which are not occurring in the gold standard dataset, and are not in the group of false positives.

Our model extracts 18 new drug-disease relations from the 203 PubMed abstracts. They are presented in Table 2, along with the extracted action, and the relative frequency of the triplet in the context of the entire analyzed corpus. The novel drug-disease pairs, shown in Table 2, are not part of the dataset we used. Because of that, we needed to manually check each pair in order to prove its correctness. We confirmed the model's predictions to be correct after searching each extracted pair in relevant biomedical literature. For instance, our model extracted a relation between vitamin D and osteoporosis (Table 2), and we can check and confirm that this pair has a proven relation, based on published papers available on the PubMed website[8].

6   Applying the DD-RDL Model on COVID-19 Related Papers

Given the global impact of the ongoing pandemic and the urgency to find solutions which will help mitigate the negative effects of COVID-19 on the public

---

[8] https://pubmed.ncbi.nlm.nih.gov/?term=%28%28Vitamin+D%29+AND+%28Osteoporosis%29%29+AND+%28receive%29.

Table 2. The novel drug-disease pairs and their actions, extracted from the PubMed abstracts. The analysis of the 203 PubMed paper abstracts resulted in the extraction of 18 novel drug-disease pairs, which we manually checked and confirmed to be correct in relevant biomedical literature.

Drug	Action	Disease	Relative frequency
Magnesium	Have	Coronary Heart Disease	0.00657
Antidiabetic	Represent	Cardiovascular Risk Factors	0.00493
Vitamin D	Receive	Osteoporosis	0.00328
Vitamin C	Suggest	Cardiovascular Disease	0.00328
Estrogen	Limit	Osteoporosis	0.00164
ACE inhibitors	Ameliorate	Diabetes Complications	0.00164
ACE inhibitors	Suggest	Cardiovascular Disease	0.00164
Atorvastatin	Contain	Cardiovascular Risk Factors	0.00164
Therapeutic Agents	Affect	Osteoporosis	0.00164
Vasoconstrictor	Stimulate	Hypertension	0.00164
Testosterone	Review	Hypogonadism	0.00164
Fluoride	Cause	Dental Fluorosis	0.00164
Vitamin D	Provide	Chronic Kidney Disease	0.00164
Calcium	Indicate	Atherosclerotic Disease	0.00164
Glucosidic Isoflavone	Prevent	Osteoporosis	0.00164
Simvastatin	Receive	Cardiovascular Disease	0.00164
Metformin	Reduce	Type 2 Diabetes	0.00164
Antidiabetic	Represent	Type 2 Diabetes	0.00164

health, we decided to apply our DD-RDL model on a set of COVID-19 related reseach papers. For that purpose, we used the CORD-19 dataset [26]. According to the results produced by our model (Table 3), we can conclude that selenium, zinc and magnesium as supplements have an effect on the maintenance of the immune system. Therefore, these supplements can be considered as positively correlated with the protection against the SARS-Cov-2 virus. On the other hand, a number of vaccines and conjugated vaccines have been developed or are in the process of development, so it is no wonder that they are very often mentioned in the abstracts. There are also protease inhibitors that, according to research, prevent the spread of the SARS-Cov-2 virus. The literature also mentions cyclohexapeptide diantin G, which occurs in the treatment of tumors. The results also indicate the appearance of copper, as new research in this field focuses on wearing masks with copper as a material that protects against the virus. We also detected that papers refer to corticosteroids as substances which can prevent the side-effects of COVID-19. Finally, the appearance of dissolved oxygen is part of every respirator, which is of great help to patients in whom this infectious disease is in its final stage. This description of each effect from the extracted drug-disease relations based on the CORD-19 dataset gives almost the same information as the daily news articles which cover the ongoing pandemic.

Table 3. The drug-action-disease triplets extracted from the CORD-19 dataset. The analysis of the abstracts resulted in the extraction of 11 triplets, which we manually checked and confirmed to be correct in relevant biomedical literature.

Drug	Action	Disease	Relative frequency
PCV13	Forestall	Community-Acquired	0.02362
Vaccine	Forestall	Influenza-like illness	0.01181
Protease Inhibitor	Depend	Irritable Bowel Syndrome	0.00393
Magnesium	Contain	Diphtheria	0.00393
Dianthin G	Block	Hypoglycemia	0.00393
Corticosteroids	Determine	Osteonecrosis of the jaw	0.00196
Dissolved Oxygen	Require	Respiratory	0.00196
Conjugate Vaccine	Present	Blood Stream Infection	0.00196
Zinc	Present	Zinc Deficiency	0.00196
Selenium	Consider	Ebola Virus Disease	0.00196
Copper	Exceed	Crohn Disease	0.00196

7   Conclusion and Future Work

This paper presented the DD-RDL model that can extract the diseases, drugs, and relations between them from medical and biological texts. In the proposed model, we apply an additional step based on SRL to discover the verbs connecting drugs and diseases. By discovering the verbs, we can add an additional label to the relation, thus allowing for the extraction of more precise knowledge from the texts. The proposed model is composed of a six-step pipeline, starting with NER, then CR, SRL as standard NLP tasks, which are then followed with the knowledge extraction steps, starting with verbs grouping, adjacency matrix construction, and at the end, the knowledge graph extraction. For all of the NLP tasks, the current state-of-the-art models based on transformers architectures are used. We evaluate the model based on a gold standard dataset for drug-disease relationships, and we demonstrate that it is possible to achieve similar results without the need of a large amount of annotated biological data or predefined semantic rules. The application of NLP in the medical domain has enormous potential. In our future work, we would like to extend the model to recognize the other important associations like disease-gene, drug-protein, and side-effects of Drugs. The second idea for future work is to implement different versions of this model for other business areas, where we will shift the focus to domain-specific knowledge extraction.

Acknowledgement. The work in this paper was partially financed by the Faculty of Computer Science and Engineering, Ss. Cyril and Methodius University in Skopje.

References

1. Abacha, A.B., Zweigenbaum, P.: Automatic extraction of semantic relations between medical entities: a rule based approach. J. Biomed. Seman. **2**(5), 1–11 (2011)
2. Aho, A.V., Corasick, M.J.: Efficient string matching: an aid to bibliographic search. Commun. ACM **18**(6), 333–340 (1975)
3. Dobreva, J., Jofche, N., Jovanovik, M., Trajanov, D.: Improving NER performance by applying text summarization on pharmaceutical articles. In: Dimitrova, V., Dimitrovski, I. (eds.) ICT Innovations 2020. CCIS, vol. 1316, pp. 87–97. Springer, Cham (2020). https://doi.org/10.1007/978-3-030-62098-1_8
4. Filannino, M., Uzuner, Ö.: Advancing the state of the art in clinical natural language processing through shared tasks. Yearbook Med. Inform. **27**(1), 184 (2018)
5. Fu, T.y., Lee, W.C., Lei, Z.: Hin2vec: explore meta-paths in heterogeneous information networks for representation learning. In: Proceedings of the 2017 ACM on Conference on Information and Knowledge Management, pp. 1797–1806 (2017)
6. Gardner, M., et al.: AllenNLP: a deep semantic natural language processing platform. In: Proceedings of Workshop for NLP Open Source Software (NLP-OSS) (2018)
7. Gottlieb, A., Stein, G.Y., Ruppin, E., Sharan, R.: Predict: a method for inferring novel drug indications with application to personalized medicine. Mole. Syst. Biol. **7**(1), 496 (2011)
8. Gu, J., Qian, L., Zhou, G.: Chemical-induced disease relation extraction with various linguistic features. Database **2016**, 042 (2016)
9. Henry, S., McInnes, B.T.: Literature based discovery: models, methods, and trends. J. Biomed. Inform. **74**, 20–32 (2017).https://doi.org/10.1016/j.jbi.2017.08.011,https://www.sciencedirect.com/science/article/pii/S1532046417301909
10. Jofche, N., Mishev, K., Stojanov, R., Jovanovik, M., Trajanov, D.: PharmKE: Knowledge extraction platform for pharmaceutical texts using transfer learning (2021)
11. Kadir, R.A., Bokharaeian, B.: Overview of biomedical relations extraction using hybrid rulebased approaches. J. Ind. Intell. Inf. **1**(3) (2013)
12. Khan, J.Y., et al.: COVID-19Base: a knowledgebase to explore biomedical entities related to COVID-19. arXiv preprint arXiv:2005.05954 (2020)
13. Kilicoglu, H., Rosemblat, G., Fiszman, M., Shin, D.: Broad-coverage biomedical relation extraction with Semrep. BMC Bioinform. **21**, 1–28 (2020)
14. Kraljevic, Z., et al.: MedCAT - Medical Concept Annotation Tool (2019)
15. Lee, K., He, L., Lewis, M., Zettlemoyer, L.: End-to-end neural co reference resolution. In: Proceedings of the 2017 Conference on Empirical Methods in Natural Language Processing, pp. 188–197. Association for Computational Linguistics, Copenhagen, Denmark, September 2017. https://doi.org/10.18653/v1/D17-1018, https://www.aclweb.org/anthology/D17-1018
16. Li, J., Sun, A., Han, J., Li, C.: A survey on deep learning for named entity recognition. CoRR abs/1812.09449 (2018), http://arxiv.org/abs/1812.09449
17. Liu, J., Abeysinghe, R., Zheng, F., Cui, L.: Pattern-based extraction of disease drug combination knowledge from biomedical literature. In: 2019 IEEE International Conference on Healthcare Informatics (ICHI), pp. 1–7. IEEE (2019)
18. Màrquez, L., Carreras, X., Litkowski, K.C., Stevenson, S.: Semantic role labeling: an introduction to the special issue. Comput. Ling. **34**, 145–159 (2008)

19. Preiss, J., Stevenson, M., Gaizauskas, R.: Exploring relation types for literature-based discovery. J. Am. Med. Inform. Assoc **22**(5), 987–992 (2015). https://doi.org/10.1093/jamia/ocv002
20. Qi, P., Zhang, Y., Zhang, Y., Bolton, J., Manning, C.D.: Stanza: a python natural language processing toolkit for many human languages. In: Proceedings of the 58th Annual Meeting of the Association for Computational Linguistics, pp. 101–108 (2020)
21. Rosario, B., Hearst, M.A.: Classifying semantic relations in bioscience texts. In: Proceedings of the 42nd Annual Meeting of the Association for Computational Linguistics (ACL-2004), pp. 430–437 (2004)
22. Sang, S., Yang, Z., Wang, L., Liu, X., Lin, H., Wang, J.: SemaTyP: a knowledge graph based literature mining method for drug discovery. BMC Bioinform. **19**(1), 1–11 (2018)
23. Shi, P., Lin, J.: Simple BERT models for relation extraction and semantic role labeling (2019)
24. Soon, W.M., Ng, H.T., Lim, D.C.Y.: A machine learning approach to coreference resolution of noun phrases. Comput. Linguist. **27**(4), 521–544 (2001). https://doi.org/10.1162/089120101753342653, https://www.aclweb.org/anthology/J01-4004
25. Torfi, A., Shirvani, R.A., Keneshloo, Y., Tavaf, N., Fox, E.A.: Natural Language Processing Advancements By Deep Learning: A Survey (2020)
26. Wang, L.L., et al.: CORD-19: The COVID-19 open research dataset (2020)
27. Wang, P., Hao, T., Yan, J., Jin, L.: Large-scale extraction of drug-disease pairs from the medical literature. J. Assoc. Inf. Sci. Technol. **68**(11), 2649–2661 (2017)
28. Wolf, T., et al.: HuggingFace's transformers: state-of-the-art natural language processing. In: Proceedings of the 2020 EMNLP (Systems Demonstrations), pp. 38–45 (2020)
29. Xia, Q., et al.: Syntax-aware neural semantic role labeling (2019)
30. Xu, R., Wang, Q.: Large-scale extraction of accurate drug-disease treatment pairs from biomedical literature for drug repurposing. BMC Bioinform. **14**(1), 1–11 (2013)
31. Xue, H., Li, J., Xie, H., Wang, Y.: Review of drug repositioning approaches and resources. Int. J. Biol. Sci. Int. J. Biol. Sci. **14**(10), 1232 (2018)
32. Yang, H., Swaminathan, R., Sharma, A., Ketkar, V., Jason, D.: Mining biomedical text towards building a quantitative food-disease-gene network. In: Learning Structure and Schemas from Documents, pp. 205–225. Springer, Cham (2011). https://doi.org/10.1007/978-3-642-22913-8
33. Zhou, R., Lu, Z., Luo, H., Xiang, J., Zeng, M., Li, M.: NEDD: a network embedding based method for predicting drug-disease associations. BMC Bioinform. **21**(13), 1–12 (2020)

Theoretical Foundations
and Information Security

Novel T-norm for Fuzzy-Rough Rule Induction Algorithm and Its Influence

Andreja Naumoski[(✉)], Georgina Mirceva, and Kosta Mitreski

Faculty of Computer Science and Engineering, Ss. Cyril and Methodius University in Skopje, Skopje, Macedonia
{andreja.naumoski,georgina.mirceva,kosta.mitreski}@finki.ukim.mk

Abstract. Machine learning algorithms can generate models in a different output form sometimes in a form of graph, time series or if-then rules. If-then rules are quite easy for humans to understand them, without any statistical understanding how the algorithm works. In this direction, the paper focuses on improving fuzzy-rough rule induction algorithms by adding a novel T-norm, particularly Einstein T-norm. The fuzzy-rough rule induction algorithm operates with two concepts (lower and upper approximation), which are very sensitive to various implicators, fuzzy tolerance relationship metrics and T-norms. In our experimental evaluation, we used four fuzzy tolerance relationship metrics, one implicator and two different T-norms: Algebraic (used previously and the one to compare) and the novel Einstein norm. The obtained experimental evaluation results revealed some interesting results, beside the improvement of the algorithm performance with the novel Einstein t-norm, the selection of fuzzy tolerance relationship sometimes can have big differences on the final output model, and on some datasets, it did not have any influence whatsoever. For future work, we plan to conduct further investigation of the influence of the implicator on the model performance.

Keywords: T-norm · Fuzzy similarity metrics · Fuzzy rough sets · Rule induction algorithms · Classification

1 Introduction

Many of the classical machine learning algorithms have advantages and disadvantages that are corrected by either correcting the algorithm itself, or introducing a totally different approach, which results in a new algorithm approach. This is the case with introduction of the fuzzy theory, and moreover the fuzzy-rough based algorithms. One implementation of this theory is used to produce rule induction models, which produce if-then rules, and in some datasets, better model accuracy. Moreover, the rule induction fuzzy-rough based algorithms can be used for solving classification and regression tasks, meaning for prediction of both nominal and numeric values. There are advantages of using the fuzzy-rough based algorithms, however there are still some drawbacks that need to be reduced. One example for this kind of problem, which is not limited only to this class of algorithms, is the curse of dimensionality, meaning that the model accuracy

© Springer Nature Switzerland AG 2022
L. Antovski and G. Armenski (Eds.): ICT Innovations 2021, CCIS 1521, pp. 115–125, 2022.
https://doi.org/10.1007/978-3-031-04206-5_9

will decrease as the number of characteristics in the initial dataset increases. One way to solve this is the implementation of pre-processing the dataset or use feature selection [1]. And this is the among the advantages of the rough set theory applicability, which allows the algorithm to preserve the inherent semantics, but also preserving generality. Furthermore, algorithms can build models that no constraints are required. This helps in the process of further automation of the rule induction process. Consequently, research in direction for combining fuzzy-rough rule induction with feature selection is more than welcome.

In the last decade [2], rough set theory is a discipline that has great interest by many researchers applied in many different problem domains. In the literature, there are few attempts to combine this theory and rule induction process to build fuzzy-rough rule induction. The research done in [3, 4] is focused on applying fuzzy theory to develop fuzzy rule set induction methods, however no fuzzy-rough concepts are considered. And since no fuzzy-rough concepts were implemented, no advantage of knowing what we know, what we do not and what we cannot know from the incomplete dataset. These are also known as three types of missing values: lost values, "do not care" condition and attribute concept values. This is achieved by introduction of lower and upper approximation and definition of the upper, lower and boundary region as stated in [5]. The fuzzy reducts concepts, combined with fuzzy-rough feature selection steps are implemented inside the method to produce rule models from the dataset. Further advancing these concerts, the researchers in [6] used fuzzy-rough methodology for rule generation algorithms. In the research made in [6], the authors used the fuzzy-rough rule induction algorithm, first to generate hierarchical decision trees, and after that convert them into rules. In [7], a method for producing the two types of rule sets is presented. There are several algorithms based on the fuzzy-rough rule induction worth mentioning, like the fuzzy algorithms described in [8, 9], alternative of association rule algorithm [10, 11], an approach for preserving semantics [12, 13] and QuickRules induction [14].

Further advancement in this area is made by introducing the approach known as VQRS (vaguely quantified rough set) algorithm [1, 15]. By introducing this change, it is possible to make the algorithm more resistant on noise and uncertainty. Since almost all real-world problem datasets exhibit these characteristics (noise and uncertainty), the application of this algorithm becomes obvious and practicality this cannot be denied. Furthermore, VQRS is sensitive to the change on the building blocks for the fuzzy-rough algorithms (similarity metrics, triangular norms and implicators). In this direction, the aim of this paper is to examine the influence of four fuzzy tolerance relationship metrics and two T-norms; namely Algebraic T-norm previously used in [16] and the novel Einstein T-norm, using several datasets with the two previously mentioned algorithms: Quick Rule and VQRS algorithm. We use AUC-ROC to assess the accuracy of the models.

The rest of this research paper is structured as follows. Section 2 provides the required details for the fuzzy-rough set theory, while Sect. 2.3 describes the VQRS algorithm. Section 3 summarizes the experimental setup and shows the results obtained using both algorithms for several classification problems, while Sect. 4 concludes the paper.

2 Fuzzy-Rough Rule Induction Algorithm

2.1 Fuzzy-Rough Rule Induction Algorithm Concepts

Before we start in details explaining the fuzzy rough rule induction algorithms and its working parts, here we first present the theory behind the fuzzy-rough set theory. So, we start by defining the fuzzy-rough analysis main concept – information system. This concept is defined as a couple (X, A), where $X = \{x_1, x_2, ..., x_n\}$ and $A = \{a_1, a_2, ..., a_n\}$ are both finite sets of objects and attributes. Both sets can have integer and nominal values, that can be real measured values or artificial to train the model. Of course, each attribute has value from a finite set of values, and the algorithm compares these values using strict equality rules. The rules are known as fuzzy tolerance relationship metrics (FTRM) and their definition is given with Eq. 1.

$$R_a(x, y) = \{(x, y) \in U^2 \mid \forall a \in B, a(x) = a(y)\} \tag{1}$$

For subsets of variables from set A, for example set B, the fuzzy B-indiscernibility relationship R_B is defined with Eq. 2, where $a \in B$ and $x, y \in X$:

$$R_B(x, y) = T(R_a(x, y)) \tag{2}$$

In Eq. 2, the T symbol represents the selected T-norm, and the B set may have different types of attributes. Now, when we define the R_B as a fuzzy tolerance metric, this will mean that it has the properties of the reflective and symmetric relationship, and therefore can be used to make approximation of the fuzzy sets in X. So, in this context, the definitions of the lower and upper approximations for fuzzy set A, R_B are given with Eq. 3 and Eq. 4.

$$\underline{R_B}(y) = \inf_{x \in X} Ip(R_B(x, y), A(x)) \tag{3}$$

$$\overline{R_B}(y) = \sup_{x \in X} T(R_B(x, y), A(x)) \tag{4}$$

Now, we have another variable in play, and that is the Implicator (Ip). An implicator Ip provide mapping $[0, 1]^2 \rightarrow [0, 1]$, which decreases in the first argument and increases in the second argument, where $Ip (1, 0) = 0$ and $Ip (0, 0) = Ip (0, 1) = Ip (1, 1) = 1$. Compared to the this implicator, with the T-norm a $[0, 1]^2 \rightarrow [0, 1]$ mapping is performed, which satisfies $T (x, 1) = x$, where x in $[0, 1]$. This mapping is an increasing, commutative and associative. As we stated previously, in our paper we will use the Einstein T-norm defined as: $(T (x, y) = ((x + y)/(1 + x * y))$ and the Algebraic T-norm $(T (x, y) = x * y)$. The KD implicator that we will use is defined as $Ip (x, y) = \max (1 - x, y)$.

All these defined concepts, by means of upper and lower approximations, now can be used for fuzzy-rough knowledge discovery processes with the concept of decision system. Based on the definition of a fuzzy-rough data analytics algorithm, a decision system defined as $(X, A \cup \{d\})$ is an information system, where d is an attribute which is excluded from the set A, known as decision. If the d attribute has numeric value, then we talk about regression problem, and if d attribute has a nominal value, then we talk about classification problem. So, based on the d values, set X is divided into several non-overlapping decision classes. To use this information, the algorithm uses decision reducts. According to the definition given in [17], the decision reduct is a subset B from A set, in which the B-positive region is a fuzzy set in X that includes object y to the level that the objects with roughly equal values for each attribute in B have equal decision values. The B-positive region as well the A-positive region can be defined as each value y and the predictive ability to calculate the dependency degree of d on B, represented with Eq. 5, according [15].

$$POS_B(y) = \left(\bigcup_{i=1}^{p} R_B\right)(y) = \max_{i=1}^{p} \inf_{x \in X} Ip(R_B(x, y), A_k(x)) \tag{5}$$

To compute the predictive power of d on attributes in B, which is a subset of attributes from set A, we will use the equation (decision reduct) shown in Eq. 6.

$$\frac{|POS_B|}{|POS_A|} = \frac{\sum\limits_{x \in X} POS_B(x)}{\sum\limits_{x \in X} POS_A(x)} \tag{6}$$

According to the definitions given in [18], the B subset keeps the decision-making power of A if degree of dependency equals to 1, and then the subset B of A is named super-reduct. If this super-reduct cannot be reduced, in fact if the positive regions in both A and B sets are equal, then the super-reduct is called decision reduct. If we want to compute all the decision-reduct for a given dataset at hand, it is considered a NP-hard problem, and therefore to implement the algorithm in practice, on many occasions it is sufficient that the algorithm generates a single decision super reduct. In this direction, many algorithms that follow this principle have been developed, but we will investigate, as we stated in the beginning of the paper, the two most widely used: Quick Reduct [1] and VQRS algorithm [15].

2.2 QuickRules Algorithm

The QuickRules algorithm [14], combines the advances of the fuzzy-rough rule induction and crisp rule induction to produce fast and reliable rule induction algorithm. As in the crisp rule induction algorithm, the QuickRules algorithm will generate rules from the training dataset, in the case of fuzzy-rough view - decision systems and then using the Quick Reduct will produce optimal model. This is done by recursively partitioning the input dataset in decision reduct, and then each class is related to a single rule. And since the QuickRules algorithm uses the partitions from the reduct process, this ensures that each class is a subset of a decision concept. We can use the attribute values to give a relevant prediction of the decision concept according [14]. So, by using the Quick Reduct concepts we can ensure that each element of the given dataset at hand will be included in the rules set that the QuickRules algorithm generated. Nevertheless, the QuickRules has some disadvantages, and this is obvious when we consider the rule generation process in which the algorithm generates rules that are too narrow. To avoid this, the rule induction step was introduced in the feature selection process in [14]. During this process, the authors in [14] made the algorithm to produce fuzzy rules to maximally cover the training objects by having only the minimum number of attributes. This combines the rule induction concept with the feature selection concept to make the QuickRules algorithm more robust and accurate.

Now, shortly we will explain the inner workings of the QuickRules algorithm. When the rule induction algorithm is been initialized, all the rule sets are set to null. Next, the rule set is kept in another set. Based on the fuzzy set, which is a subset of the main set, the present coverage degree for each entity is calculated by the existing set of rules. To achieve this, a covered(C) function is used, in which C is the fuzzy set, that will return the set of objects that are greatly contained in C. The covered function is specified as:

$$covered(C) = \{x \in X \,|\, Cov(x) = POS_A(x)\} \tag{7}$$

So, if we examine Eq. 7, an object of interest is considered by the algorithm and it will be included into the final set, if its membership function is equal to the positive region of the entire attribute set, according to the authors in [14]. It is vital to note here that a rule of any object is considered only when it has not been covered. The same applies to any of the attributes in set $B \cup \{a\}$, and this will mean that the object of interest should belong maximally to $POS_{B \cup a}$. The algorithm produces a rule for every object of interest, and this rule will be contained in the final set, when the tolerance class for that object is fully included in the decision concept. Based on this, the tolerance class of a given attribute is a good indicator of the concept. Furthermore, to ensure that the final set of rules is informative and compact, the algorithm checks if the newly created rules are only added if the coverage of the new rule has a higher value of the existing rule in the set. When the current rule has lower coverage compared to the new rule, then the current rule is removed. At the end of the rule generation process, if any new rule fulfils the given condition, the coverage is included in the set. If a certain object is fully covered, then new rules are not added. After all the objects from the input dataset are tested and they cover all the positive regions, the rule induction process is terminated. As a final output of the QuickRules algorithm is a rule set that covers all the objects of interest.

2.3 VQRS Algorithm

The second algorithm that we will describe is the VQRS algorithm, which in fact uses vague quantifiers to revise the hybridization process for the definition of the two approximations. This is achieved by using external qualifiers for upper approximations and universal quantifiers for lower approximation. The object from the dataset at hand belongs to the upper approximation when is associated to an element that belongs in the dataset, and in contrast of this, the lower approximations only contain objects that are related to all the elements in the dataset. This also makes the VQRS algorithm more robust when dealing with errors, noise, uncertainty, and inconsistency, which are properties of a real dataset. This is because the VQRS algorithm [15] relates the definition of the two approximations with vague quantifiers concept and allows the algorithm to deal with these real-world properties of the dataset.

Before we proceed with experimental setup, we define the vague quantifiers. A vague quantifier is used to soften the two approximations, so y belongs to the upper approximation to the extent of *some* elements, and y belongs to the lower approximation to the extent the most elements in A, according to [15]. The object itself can belong to the two approximations and the VQRS uses vague quantifiers to determine in which set it belongs. To do this, VQRS algorithm, defines a set of two elements Q_1 and Q_2 to represent the fuzzy quantifiers, and then the two approximations of A by R_a [15], defined with Eq. 8 and 9.

$$\underline{R_{Q1}}(y) = Q_1\left(\frac{|R_B \cap A|}{R_B}\right) = Q_1\left(\frac{\sum\limits_{x \in X} \min(R_A(x, y), A(x))}{\sum\limits_{x \in X} R_A(x, y)}\right) \tag{8}$$

$$\overline{R_{Q2}}(y) = Q_2\left(\frac{|R_B \cap A|}{R_B}\right) = Q_2\left(\frac{\sum\limits_{x \in X} \min(R_B(x, y), A(x))}{\sum\limits_{x \in X} R_B(x, y)}\right). \tag{9}$$

In Eq. 8 and 9, the fuzzy set intersection is a combination of min operation (T-norm as a first element and fuzzy set cardinality as a second element), and this is done by sigma-count operation defined by [15]. To calculate the $R_A(x, y)$ part of the equation, in our experiments we use four different fuzzy tolerance relationship metrics (FTRM). The numeric values of a are real numbers, who are in a closed interval and compared using FTRM. In our experimental setup, which will be explained later in more details, four FTRM metrics are used.

$$FTRM_1 = 1 - \frac{|a(x) - a(y)|}{|a_{\max} - a_{\min}|} \tag{10}$$

$$FTRM_2 = \exp\left(-\frac{(a(x) - a(y))^2}{2\sigma_a^2}\right) \tag{11}$$

$$FTRM_3 = \max\left(\min\left(\frac{(a(y) - (a(x) - \sigma_a))}{(a(x) - (a(x) - \sigma_a))}, \frac{((a(x) + \sigma_a) - a(y))}{((a(x) + \sigma_a) - a(x))}\right), 0\right) \tag{12}$$

$$FTRM_4 = \max\left(0, \min\left(1, \frac{\beta - \alpha * |a(x) - a(y)|}{a_{\max} - a_{\min}}\right)\right), \quad \alpha = 0.5, \beta = 1.0 \tag{13}$$

The standard deviation σ_a is denoted for each a attribute value, while x and y variables are elements in the X set. The four FTRM represented with Eq. 10–Eq. 13, are considered in the experiments to test the impact of the newly introduced T-norm against the previously used T-norm, in both fuzzy-rough rule induction algorithms.

3   Experimental Setup and Results

3.1   Experimental Setup

In this section, the VQRS fuzzy-rough rule induction algorithm settings parameters are presented as well as short description over fourteen datasets from [16, 18]. A detailed

Table 1. Characteristics of the datasets that are used in the experiments.

Dataset	Instances	Attributes	Dataset	Instances	Attributes
Breast-cancer	569	32	Vehicle	946	19
Credit-a	690	15	EcoData1	218	117
Credit-g	1000	20	EcoData2	218	117
Glass	214	10	EcoData3	218	117
Heart-Statlog	270	14	EcoData4	218	117
Ionosphere	351	35	EcoData5	218	117
Iris	150	5	Wine	178	14

description for each of the datasets that are used to evaluate certain aspects of this algorithm are presented in Table 1.

As we mentioned, we use two T-norms: The novel Einstein T-norm and the Algebraic T-norm for evaluation, as well as the KD implicator for all experiments. All these operators were defined in the previous section. The VQRS fuzzy-rough rule induction algorithm parameter settings are set to: $Q(\alpha, \beta) = Q(0.2; 1.0)$, according to Eq. 14. The rest of the algorithm parameter settings are set to their default values.

$$Q_{(\alpha,\beta)}(x) = \begin{cases} 0, & x \leq \alpha \\ \frac{2(x-\alpha)^2}{(\beta-\alpha)^2}, & \alpha \leq x \leq \frac{\alpha+\beta}{2} \\ 1 - \frac{2(x-\alpha)^2}{(\beta-\alpha)^2}, & \frac{\alpha+\beta}{2} \leq x \leq \beta \\ 1, & \beta \leq x \end{cases} \tag{14}$$

For evaluation of the models obtained using the novel T-norm metric, we use the standard 10-fold cross validation procedure. The results from the experimental evaluation are presented in Table 2 for descriptive/predictive performance of the new T-norm and how this T-norm performed over 14 datasets with QuickRules algorithm, and Table 4 with VQRS algorithm. Later, in Table 3 and 5 we present the difference between the results obtained with the new Einstein T-norm and the previously used Algebraic T-norm over the 14 datasets with QuickRules/VQRS algorithm accordingly.

3.2 Experimental Results

In this section we present and discuss the experimental results. In the tables, beside the names of datasets that are used for evaluation, we evaluate the fuzzy tolerance relationship metrics that contain four elements {FTRM$_1$(Eq. 10), FTRM$_2$ (Eq. 11), FTRM$_3$ (Eq. 12), FTRM$_4$ (Eq. 13)}, presented as Train/Test$^{\text{FTR metric}}$. AUC_ROC metric is used for both descriptive (Train) and predictive (Test) model performance.

In Tables 2 and 3, experimental evaluation of the QuickRules algorithm is presented, while the experimental results from the VQRS algorithm are presented in Tables 4 and 5. The results from Table 2 clearly indicate that in many of the datasets, the descriptive and predictive accuracy is above 0.8. Exceptions from this rule are several datasets

Table 2. Results of classification task evaluation using AUC_ROC metric for models from Quick-Rules algorithm. Bolded results present the AUC-ROC value above 0.8 for both training and testing model performance.

Dataset	Train^1	Test^1	Train^2	Test^2	Train^3	Test^3	Train^4	Test^4
Breast-cancer	**0.98**	0.55	**0.98**	0.55	**0.98**	0.55	**0.98**	0.55
Credit-a	**0.95**	**0.81**	**0.98**	**0.85**	**1.00**	**0.83**	**0.95**	**0.81**
Credit-g	**1.00**	0.66	**1.00**	0.64	**1.00**	0.67	**1.00**	0.66
Glass	**0.82**	0.77	**0.90**	**0.81**	**1.00**	**0.85**	**0.82**	0.77
Heart-Statlog	**0.98**	**0.89**	**1.00**	**0.89**	**1.00**	**0.80**	**0.98**	**0.89**
Ionosphere	**1.00**	**0.96**	**1.00**	**0.98**	**1.00**	**0.91**	**1.00**	**0.96**
Iris	**0.99**	**0.98**	**0.99**	**0.99**	**1.00**	**1.00**	**0.99**	**0.98**
Vehicle	**0.93**	**0.89**	**0.98**	**0.91**	**1.00**	**0.88**	**0.93**	**0.89**
EcoData1	**0.98**	0.48	**0.99**	0.63	**0.93**	0.57	**0.98**	0.48
EcoData2	**0.97**	0.53	**0.98**	0.60	**0.93**	0.56	**0.97**	0.53
EcoData3	**0.93**	0.53	**0.98**	0.58	**0.94**	0.51	**0.93**	0.53
EcoData4	**0.99**	0.44	**0.99**	0.42	**0.93**	0.44	**0.99**	0.44
EcoData5	**0.97**	0.48	**0.99**	0.59	**0.92**	0.53	**0.97**	0.48
Wine	**1.00**	**1.00**	**1.00**	**1.00**	**1.00**	**0.99**	**1.00**	**1.00**

(Breast-cancer, credit-g and EcoData series of datasets), and according to the results, the models from these algorithms are over-fitted. The overall conclusion is that the Quick-Rules algorithm achieves satisfactory results for most of the datasets. If we compare these results, with the previously published results [16], when we use Algebraic T-norm (see Table 3), we note that for most of the datasets, there is no significant increase or decrease of the model performance. If we closely analyse the results from Table 3, this range of model performance is between −0.03 to 0.03. There is no difference of model performance among the FTR metrics used in Table 2 and Table 3.

Next, we analyse the impact on the novel Einstein T-norm on the VQRS algorithm. The results from the experimental evaluation among the four FTR metrics are presented in Table 4.

As in Table 2, here the descriptive model performance in all the cases is above 0.8, most of them are close to 0.95, however, the experimental evaluation of the models on unseen data reveals a similar picture as the results presented for the QuickRules algorithm. Many of the models produced AUC-ROC values greater than 0.8, however the same set of datasets (Breast-cancer, credit-g and EcoData series of datasets) achieved model results between 0.4 and 0.7. For FTRM$_1$ and FTRM$_4$ metrics, a small difference is noted compared to the other two metrics. The descriptive models of FTRM$_1$ and FTRM$_4$ performed slightly worse than the model performance of the Algebraic T-norm, while the predictive performance obtained better performance.

Table 3. Results of classification task evaluation using AUC-ROC metric (difference between T-norm Einstein and T-norm Algebraic) for the QuickRules algorithm. Bolded results denote best performing model for each dataset, while underlined results denote worst performance for each dataset.

Dataset	Train[1]	Test[1]	Train[2]	Test[2]	Train[3]	Test[3]	Train[4]	Test[4]
Breast-cancer	0.00	0.00	0.00	0.00	0.00	0.00	0.00	0.00
Credit-a	0.00	0.00	−0.01	**0.02**	0.00	0.00	0.00	0.00
Credit-g	0.00	0.00	0.00	0.00	0.00	0.00	0.00	0.00
Glass	**0.01**	**0.03**	0.01	**0.01**	0.00	**0.01**	**0.01**	**0.03**
Heart-Statlog	**0.01**	0.00	0.00	0.00	0.00	**0.01**	**0.01**	0.00
Ionosphere	0.00	**0.01**	0.00	**0.01**	0.00	0.00	0.00	**0.01**
Iris	0.00	**0.01**	0.00	0.00	0.00	0.01	0.00	**0.01**
Vehicle	**0.02**	**0.01**	**0.02**	0.00	0.00	0.00	**0.02**	**0.01**
EcoData1	0.00	0.00	0.00	**0.03**	0.00	0.00	0.00	0.00
EcoData2	**0.01**	0.00	0.00	0.00	0.00	0.00	**0.01**	0.00
EcoData3	−0.03	0.00	0.00	0.00	0.00	0.00	−0.03	0.00
EcoData4	0.00	**0.01**	0.00	−0.01	0.00	−0.01	0.00	**0.01**
EcoData5	−0.01	0.00	0.00	**0.02**	0.00	0.00	−0.01	0.00
Wine	0.00	0.00	0.00	0.00	0.00	0.00	0.00	0.00

Table 4. Results of classification task evaluation using AUC_ROC metric for models from VQRS algorithm. Bolded results present the AUC-ROC value above 0.8 for both training and testing model performance.

Dataset	Train[1]	Test[1]	Train[2]	Test[2]	Train[3]	Test[3]	Train[4]	Test[4]
Breast-cancer	**0.95**	0.56	**0.95**	0.56	**0.95**	0.56	**0.95**	0.56
Credit-a	**0.96**	**0.82**	**1.00**	**0.85**	**0.96**	**0.86**	**0.96**	**0.82**
Credit-g	**1.00**	0.62	**1.00**	0.62	**1.00**	0.64	**1.00**	0.62
Glass	0.68	0.72	**0.99**	**0.80**	0.68	0.86	0.68	0.72
Heart-Statlog	**0.96**	**0.83**	**1.00**	**0.88**	**0.96**	**0.81**	**0.96**	**0.83**
Ionosphere	**1.00**	**0.96**	**1.00**	**0.98**	**1.00**	**0.94**	**1.00**	**0.96**
Iris	**0.99**	**0.93**	**1.00**	**0.97**	**0.99**	**0.99**	**0.99**	**0.93**
Vehicle	**0.85**	**0.84**	**1.00**	**0.90**	**0.85**	**0.89**	**0.85**	**0.84**
EcoData1	**0.97**	0.67	**0.92**	0.67	**0.97**	0.58	**0.97**	0.67
EcoData2	**0.98**	0.63	**0.89**	0.59	**0.98**	0.50	**0.98**	0.63
EcoData3	**0.95**	0.59	**0.93**	0.56	**0.95**	0.51	**0.95**	0.59
EcoData4	**0.96**	0.48	**0.92**	0.34	**0.96**	0.48	**0.96**	0.48
EcoData5	**0.96**	0.59	**0.93**	0.55	**0.96**	0.50	**0.96**	0.59
Wine	**1.00**	**1.00**	**1.00**	**1.00**	**1.00**	**0.99**	**1.00**	**1.00**

Table 5. Results of classification task evaluation using AUC-ROC metric (difference between T-norm Einstein and T-norm Algebraic) for the VQRS algorithm. Bolded results denote best performing model for each dataset, while underlined results denote worst performance for each dataset.

Dataset	Train[1]	Test[1]	Train[2]	Test[2]	Train[3]	Test[3]	Train[4]	Test[4]
Breast-cancer	0.00	0.00	0.00	0.00	0.00	0.00	0.00	0.00
Credit-a	0.00	0.00	0.00	0.00	0.00	0.01	0.00	0.00
Credit-g	−0.03	0.00	0.00	0.00	0.00	0.00	0.00	0.00
Glass	**0.02**	**0.05**	0.00	0.00	**0.01**	**0.02**	−0.03	**0.05**
Heart-Statlog	0.00	0.00	0.00	0.00	0.00	0.00	**0.02**	0.00
Ionosphere	**0.05**	−0.01	0.00	**0.01**	0.00	**0.01**	0.00	−0.01
Iris	**0.10**	−0.01	0.00	**0.01**	0.00	0.00	**0.05**	−0.01
Vehicle	**0.01**	**0.07**	**0.03**	**0.02**	0.00	0.00	**0.10**	**0.07**
EcoData1	0.00	**0.02**	−0.02	−0.01	−0.04	−0.01	**0.01**	**0.02**
EcoData2	−0.02	**0.05**	0.00	**0.02**	0.00	−0.01	0.00	**0.05**
EcoData3	**0.01**	**0.02**	**0.01**	−0.01	0.00	−0.01	−0.02	**0.02**
EcoData4	**0.01**	**0.07**	**0.02**	−0.07	0.00	−0.01	**0.01**	**0.07**
EcoData5	0.00	0.00	−0.08	−0.01	0.00	−0.01	**0.01**	0.00
Wine	0.00	0.00	0.00	0.00	0.00	0.00	0.00	0.00

And finally, we discuss the results from the model evaluation comparison between the previously used Algebraic T-norm and Einstein T-norm for the VQRS algorithm. By comparing the results, we noted that the difference between Einstein T-norm and Algebraic T-norm model performance ranges between −0.1 and 0.1. Here we noted a difference between the FTR metrics applied with the VQRS algorithm, in which FTRM$_1$ and FTRM$_4$ models obtain higher results than FTRM$_2$ and FTRM$_3$.

4   Conclusion

In this research paper, we introduced a novel T-norm, particularly the Einstein T-norm for QuickRules and VQRS algorithms. We have also analyzed the influence of Einstein T-norm in comparison with the Algebraic T-norm in combination with four FTRM. As we noted in our paper, these are a fuzzy-rough algorithms, with fuzzy-rough concepts (defining upper, lower and boundary regions) and able to cope with the noise and uncertainty inside more efficiently on the provided datasets. This was put on test, and from the experimental results presented in this paper, we can note that Einstein T-norm in combination with FTRM$_1$ and FTRM$_4$ metric produced models with slightly better model performance compared with the Algebraic T-norm. And this is very interesting, which invokes more research, suggesting that maybe the implicator metric is the one that needs to be considered to increase their predictive power performance.

Acknowledgement. This work was partially financed by the Faculty of Computer Science and Engineering at the Ss. Cyril and Methodius University in Skopje.

References

1. Jensen, R., Cornelis, C., Shen, Q.: Hybrid fuzzy-rough rule induction and feature selection. In: IEEE International Conference on Fuzzy Systems, 2009. FUZZY-IEEE 2009, pp. 1151–1156 (2009)
2. Pawlak, Z.: Rough Sets: Theoretical Aspects of Reasoning About Data. Kluwer Academic Publishing (1991)
3. Hsieh, N.-C.: Rule extraction with rough-fuzzy hybridization method. In: Washio, T., Suzuki, E., Ting, K.M., Inokuchi, A. (eds.) PAKDD 2008. LNCS (LNAI), vol. 5012, pp. 890–895. Springer, Heidelberg (2008). https://doi.org/10.1007/978-3-540-68125-0_89
4. Shen, Q., Chouchoulas, A.: A rough-fuzzy approach for generating classification rules. Patt. Recogn. **35**(11), 2425–2438 (2002)
5. Greco, S., Inuiguchi, M., Slowinski, R.: Fuzzy rough sets and multiple-premise gradual decision rules. Int. J. Approximate Reasoning **41**, 179–211 (2005)
6. Wang, X., Tsang, E.C.C., Zhao, S., Chen, D., Yeung, D.S.: Learning fuzzy rules from fuzzy samples based on rough set technique. Inf. Sci. **177**(20), 4493–4514 (2007)
7. Hong, T.P., Liou, Y.L., Wang, S.L.: Learning with hierarchical quantitative attributes by fuzzy rough Sets. In: Proceedings of the Joint Conference on Information Sciences. Advances in Intelligent Systems Research (2006)
8. Drobics, M., Bodenhofer, U., Klement, E.P.: FS-FOIL: an inductive learning method for extracting interpretable fuzzy descriptions. Int. J. Approx. Reason **32**, 131–152 (2003)
9. Prade, H., Richard, G., Serrurier, M.: Enriching relational learning with fuzzy predicates. In: Proceedings of Principles and Practice of Knowledge Discovery in Databases, pp. 399–410 (2003)
10. Cloete, I., Van Zyl, J.: Fuzzy rule induction in a set covering framework. IEEE Trans. Fuzzy Syst. **14**(1), 93–110 (2006)
11. Xie, D.: Fuzzy associated rules discovered on effective reduced database algorithm. Proceedings of the IEEE International Conference on Fuzzy Systems, 779–784 (2005)
12. Marin-Blazquez, J.G., Shen, Q.: From approximative to descriptive fuzzy classifiers. IEEE Trans. Fuzzy Syst. **10**(4), 484–497 (2002)
13. Qin, Z., Lawry, J.: LFOIL: linguistic rule induction in the label semantics framework. Fuzzy Sets Syst. **159**(4), 435–448 (2008)
14. Jensen, R., Cornelis, C.: A new approach to fuzzy-rough nearest neighbour classification. In: Chan, C.-C., Grzymala-Busse, J.W., Ziarko, W.P. (eds.) RSCTC 2008. LNCS (LNAI), vol. 5306, pp. 310–319. Springer, Heidelberg (2008). https://doi.org/10.1007/978-3-540-88425-5_32
15. Cornelis, C., De Cock, M., Radzikowska, A.M.: Vaguely quantified rough sets. In: An, A., Stefanowski, J., Ramanna, S., Butz, C.J., Pedrycz, W., Wang, G. (eds.) RSFDGrC 2007. LNCS (LNAI), vol. 4482, pp. 87–94. Springer, Heidelberg (2007). https://doi.org/10.1007/978-3-540-72530-5_10
16. Naumoski, A., Mirceva, G., Mitreski, K.: Influence of algebraic t-norm on different indiscernibility relationships in fuzzy-rough rule induction algorithms. In: Trajanov, D., Bakeva, V. (eds.) ICT Innovations 2017. CCIS, vol. 778, pp. 120–129. Springer, Cham (2017). https://doi.org/10.1007/978-3-319-67597-8_12
17. Jensen, R., Shen, Q.: Computational Intelligence and Feature Selection: Rough and Fuzzy Approaches. Wiley-IEEE Press, USA (2008)
18. Blake, C.L., Merz, C.J.: UCI Repository of Machine Learning Databases. University of California, Irvine (1998). http://archive.ics.uci.edu/ml/

De Bruijn-Based and k-Ary n-Cube-Based Algebraic Models in Fog Environments

Pedro Juan Roig[1]([⊠])[iD], Salvador Alcaraz[1][iD], Katja Gilly[1][iD],
Cristina Bernad[1][iD], and Sonja Filiposka[2][iD]

[1] Miguel Hernández University (Elche), Alicante, Spain
{proig,salcaraz,katya,cbernad}@umh.es
[2] Ss. Cyril and Methodius University (Skopje), Skopje, North Macedonia
sonja.filiposka@finki.ukim.mk

Abstract. In order to deal with offloading tasks, fog computing involves
a set of distributed hosts, being interconnected according to a certain
switching topology, upon which overlay networks may be configured.
These type of logical networks will be adapted to fit the needs of different
kinds of application flows, resulting in diverse sorts of traffic, each one
with its own constraints regarding quality of service. In this context,
de Bruijn graphs are reviewed as a feasible technology to be employed
for traffic forwarding, and furthermore, k-ary n-cubes are also presented
as an alternative option. Additionally, the role of de Bruijn shapes for
dealing with position location in different dimensions is also discussed.

Keywords: De Bruijn graph · De Bruijn hypertorus · De Bruijn
torus · Fog computing · k-ary n-cube

1 Introduction

Fog computing is an extension of cloud computing paradigm, where remote
resources are no longer located in a further cloud, but instead, they are located
right on the edge of the network, hence providing interesting benefits when dealing with low-latency applications or low-resourced devices [1].

Fog ecosystems are ideal to work with IoT devices, as those are resource-
constraint items regarding computer processing, storage, RAM memory, band-
width and power supply, which makes them often times unfit to work with remote
cloud deployments due to latency and jitter issues [2].

Special attention may be paid to moving IoT devices, as they might be moving around the coverage area of a fog domain, and because of its constrained
features, they may need to have their computing resources as close as possible
at all times in order to maximise its performance. This may well be achieved
by undertaking the migration of remote resources towards the host being the
closest to where the moving IoT device is located in a given time [3].

© Springer Nature Switzerland AG 2022
L. Antovski and G. Armenski (Eds.): ICT Innovations 2021, CCIS 1521, pp. 126–141, 2022.
https://doi.org/10.1007/978-3-031-04206-5_10

It is to be said that those remote computing resources are deployed as different entities, such as virtual machines (VM) or docker containers, even though their live migration processes allow for the same techniques. Regarding the literature, there are far more references to the VM's live migration than containers' [4].

On the other hand, it is to be noted that fog environments are growing by the day, as they include both more users and more coverage areas, which in turn, lead to much more applications with different needs of resources [5]. In order to cope with such a diverse amount of requirements, the utilisation of logical overlay networks permits to abstract away from the real network designs [6], also called underlay networks, so as to be able to adapt to the constraints of each traffic flow [7].

This way, dynamic overlay networks have been lately benefited due to the increase in dynamic peer-to-peer (P2P) networks for finding and working with distributed data objects, the increase in virtual private networks (VPN) for business-related purposes, the rise in multicast content distribution of different types of streams, the constant development of new high-demanding applications where each subscriber requires its own constraints regarding quality of service, making for service overlay networks, and the implementation of techniques to monitor and act on the status and quality of internet paths, making for resilient overlay networks [8].

In this context, there are different topologies related to interconnection networks so as to construct overlay networks, even though de Bruijn graphs and k-ary n-cubes are among the most interesting due to its regular structure, which get better results than irregular graphs, and are easier to deal with than complete graphs, thus bringing up better results. Regarding scalable QoS-constrained overlay networks, it may seem that the latter gets higher route redundancy, whereas the former gets shorter distances in average between random pairs of nodes [9].

The organisation of the rest of this paper goes as follows: first, Sect. 2 exposes the de Bruijn sequencies, next, Sect. 3 introduces the de Bruijn graphs, then, Sect. 4 presents the Wong algorithm, afterwards, Sect. 5 proposes the de Bruijn graph as a forwarding scheme, in turn, Sect. 6 puts forward the de Bruijn tori, after that, Sect. 7 brings about the de Bruijn hypertori, later on, Sect. 8 exhibits the k-ary n-cubes, then, Sect. 9 proposes the k-ary n-cubes as a forwarding scheme, next, Sect. 10 suggests the use of de Bruijn shapes as location schemes, and eventually, Sect. 11 draws some final conclusions.

2   De Bruijn Sequence

In 1946, de Bruijn research was about strings of integers of a certain length covering all possible substrings of a given length on a non-repeating basis [10]. Those sequences may be seen as objects related to combinatorics, basically to the field of graph theory. Each string exhibits the shortest cycle of digits of length n in alphabet k, with $k \geq 2$ and $n \geq 1$, where each particular substring appears exactly once [11].

Such cyclic strings are known as de Bruijn sequences, and in its honour, they are denoted as $B(k, n)$. It is to be remarked that they have a lot of applications in location detection, angle encoding, cryptography, pseudorandom arrays and sequences, coding theory or key-lock breaking [12].

In this sense, special attention ought to be paid to the decoding problem, meaning the discovery of the position of a specific substring within a given cyclic string [13], which is not a trivial issue. In a certain way, this may be seen as the opposite of the construction problem, that being the process of building up such cyclic strings out of a given set of substrings, which is basically easier.

Anyway, the structure of a cyclic sequence may be appreciated in Fig. 1, where each node has a link to its predecessor and another link to its successor, considering that the nodes located on both edges have a wraparound link joining them both together [14].

Fig. 1. One-dimensional toroidal array (4 nodes).

There are distinct instances for each de Bruijn sequence, although those obtained by simple rotations on a particular instance are not to be considered as different solutions. Hence, circular shifting does not create new instances.

On the other hand, it is to be said that reflections on a particular instance are to be taken into account as different solutions, hence two diverse sequences are to be considered in each cycle, such as the clockwise and the counterclockwise. Therefore, flipping around does create new instances.

Specifically, the number of overall de Bruijn sequences for determined values of k-alphabet and length n is given in expression (1), where the length of each of those instances regarding particular values for k and n is given by k^n.

$$\frac{(k!)^{k^{n-1}}}{k^n} \tag{1}$$

As an example, Fig. 2 exhibits a de Bruijn sequence for a binary alphabet ($k = 2$) and three-length words ($n = 3$), where it may be appreciated that each substring of length 3 appears precisely once in the sequence of length $2^3 = 8$. Additionally, there are up to $2^{2^2}/2^3 = 2$ distinct instances, such as 00010111 and 11101000, which specifically account for the reflection of each other, as it may be depicted in both the linear and the circular representations of the former string.

Fig. 2. De Bruijn sequence with $k = 2$ and $n = 3$, shown in linear and circular form.

As a side note, here it is a particular instance of each of the de Bruijn sequences whose length is up to 64 digits:

Table 1. De Bruijn sequences whose length is up to 64.

B(k,n)	Items	An instance of that de Bruijn sequence
$\{0,1\}$	$B(2,2)$	0011
$\{0,1\}$	$B(2,3)$	00010111
$\{0,1\}$	$B(2,4)$	0000100110101111
$\{0,1\}$	$B(2,5)$	00001010111011000111110011010010
$\{0,1\}$	$B(2,6)$	0000001000011000101000111001001011001101001111010101110110111111
$\{0\cdots2\}$	$B(3,2)$	001122102
$\{0\cdots2\}$	$B(3,3)$	000111212220101200202102211
$\{0\cdots3\}$	$B(4,2)$	0011223302103132
$\{0\cdots3\}$	$B(4,3)$	0001110100202221211220030333232233131133013203210310231201230213
$\{0\cdots4\}$	$B(5,2)$	00102112041422430332313440
$\{0\cdots5\}$	$B(6,2)$	00102030405112131415223242530334354455
$\{0\cdots6\}$	$B(7,2)$	0010211204130614031505516225235343644246332654560
$\{0\cdots7\}$	$B(8,2)$	001020304050607112131415161722324252627334353637445464755657667

3 De Bruijn Graph

The main feature in de Bruijn sequences is that a given string of length n shares its $n-1$ leftmost digits with its predecessor (being the $n-1$ rightmost digits of this neighbour), whereas it also shares its $n-1$ rightmost digits with its successor (being the $n-1$ leftmost digits of that neighbour), hence the overlapping of both $n-1$ substrings may bring up an n string, and clearly, the concatenation of all n strings may bring about a k^n sequence. In other words, taking the rightmost symbol of each unique string in an ordered manner will result in the de Bruijn sequence. This procedure may allow to build up this sort of sequences by two methods, such as a graphical one and an algorithmic one.

The classical way to construct them is through de Bruijn graphs, which are directed graphs being regular, connected, cyclic and simple (even though loopback links arises for values with a single digit). Therefore, a Hamiltonian path along this graph may allow to go through all k^n nodes just once, and in turn, getting back to the initial node, thus obtaining a de Bruijn sequence $B(k,n)$. On the other hand, a Eulerian path along that graph may permit to go through all k^{n+1} edges only once, and then, getting back to the initial node, thus achieving a de Bruijn sequence $B(k, n+1)$, where each string results out of the concatenation of each node with its outgoing edge according to this path.

As an example, Fig. 3 shows the de Bruijn graph for a binary alphabet and three-word nodes, which allows the construction of both $B(2,3)$ going through all nodes exactly once and $B(2,4)$ moving through all edges precisely once.

In the former, each hop might be seen as applying the value of the outgoing edge as a left shift to the departing node, whilst in the latter, the value of the outgoing edge is concatenated to the departing node.

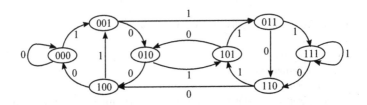

Fig. 3. Binary de Bruijn graph: Hamiltonian $B(2,3)$ and Eulerian $B(2,4)$.

Likewise, Fig. 4 exhibits a trefoil de Bruijn graph for a 3-alphabet and two-word nodes, leading to build up both $B(3,2)$ by means of a Hamiltonian path and $B(3,3)$ by taking a Eulerian path.

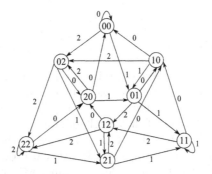

Fig. 4. Ternary de Bruijn graph: Hamiltonian path B(3,2) and Eulerian path B(3,3).

4   Wong Algorithm

Alternatively, different algorithms have been proposed in the literature so as to achieve de Bruijn sequences without using de Bruijn graphs. However, the most efficient algorithm was designed by Wong [15], which is able to get a de Bruijn sequence in $O(1)$-amortised time per digit for any k-alphabet.

The version for a binary alphabet is pretty easy and it only involves checking if a particular string is a necklace, that being defined as the lexicographically earliest string in an equivalence class of strings under rotation [16], which is not to be confused with a bracelet, which involves both rotation and reflection. Therefore, the following string is given by expression (2), where $k = \{0,1\}$.

$$f_2(b_1 b_2 \cdots b_n) = \begin{cases} b_2 b_3 \cdots b_n \overline{b_1} & \text{if } b_2 b_3 \cdots b_n b_1 \text{ is a necklace;} \\ b_2 b_3 \cdots b_n b_1 & \text{otherwise.} \end{cases} \quad (2)$$

On the other hand, the version for a generic k-alphabet is quite similar, although it requires a bit more checking, and it is given by expression (3), where alphabet $k = \{0 \cdots k - 1\}$ and b happens to be the highest value in $a_2 \cdots a_n b$ not being a necklace, or 0 instead.

$$f_k(a_1 a_2 \cdots a_n) = \begin{cases} a_2 a_3 \cdots a_n b & \text{if } a_1 = k - 1; \\ a_2 a_3 \cdots a_n (a_1 + 1) & \text{if } a_1 \neq k - 1 \text{ and} \\ & a_2 a_3 \cdots a_n (a_1 + 1) \text{ is a necklace;} \\ a_2 a_3 \cdots a_n a_1 & \text{otherwise.} \end{cases} \quad (3)$$

As an example, Table 2 shows how to get the different $k^n = 8$ nodes for $B(2,3)$ in a correct order, starting at node 000, where the most significant bit of each node makes up the corresponding de Bruijn sequence. Likewise, Table 3 does it to the $k^n = 9$ nodes for $B(3,2)$, starting at node 00.

Table 2. Wong algorithm to achieve de Bruijn sequence $B(2,3)$.

Now: $b_1 b_2 b_3$	Test: $b_2 b_3 1$	Necklace?	Next: YES: $b_2 b_3 \overline{b_1}$ NO: $b_2 b_3 b_1$
000	001	Yes	001
001	011	Yes	011
011	111	Yes	111
111	111	Yes	110
110	101	No	101
101	011	Yes	010
010	101	No	100
100	001	Yes	000

5 De Bruijn Graph as a Forwarding Scheme

In order to carry out a forwarding scheme for overlay networks based on de Bruijn graphs, the first thing to do is to set up the inwards and outwards directions for a generic switch i, as exposed in Fig. 5. This picture exhibits up to k links coming in from other nodes and other k links going out to other nodes, which may allow to design a fault-tolerant forwarding strategy to bypass failed nodes.

This design may grant benefits related to scalability, resilience and load balancing in fog ecosystems with heterogeneous network devices (smartphones, tablets, raspberry pi, smart vehicles) consuming high traffic rates (such as video streams) by decoupling the network architecture between a tight macro level,

Table 3. Wong algorithm to achieve de Bruijn sequence $B(3, 2)$.

Now: $a_1 a_2$	$a_1 = 2$?	$a_2 b$	Necklace?	$a_2(a_1 + 1)$	Necklace?	Next string:
00	No	-	-	01	Yes	01
01	No	-	-	11	Yes	11
11	No	-	-	12	Yes	12
12	No	-	-	22	Yes	22
22	Yes	21	No	-	-	21
21	Yes	10	No	-	-	10
10	No	-	-	02	Yes	02
02	No	-	-	21	No	20
20	Yes	00	Yes	-	-	00

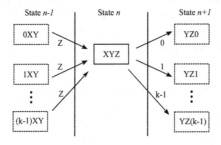

Fig. 5. Inwards and Outwards paths in a de Bruijn node.

where the symmetrical features of de Bruijn graphs may allow for easy and stable forwarding schemes, and a loose micro level, where a high degree of freedom in neighbour selection may permit to enhance the aforesaid characteristics [17].

Regarding the port layout for a generic switch, it may allow a number M of links for local hosts (from 0 to $M - 1$), whilst the next k links ($M \cdots M + k - 1$) are outgoing and the following k links are incoming, as depicted in Fig. 6.

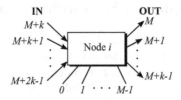

Fig. 6. De Bruijn topology with ports.

Taking all that into account, it is to be noted that if a message is received at any port p, the first thing to be checked out is whether its destination b is one of its local hosts $i \times M + p$. If that is the case, such a message is sent through its downlink port connecting to destination, or otherwise, it is to be found out its longest rightmost substring in the identifier of this node i matching the leftmost counterpart in the destination switch, given by $\lfloor b/M \rfloor$. Afterwards, it is examined whether the node corresponding to inserting the next symbol after the aforesaid substring in the latter to the rightmost position of the former is available, and if so, the message is forwarded on to that node, or otherwise, it is done towards the node corresponding to the result of inserting the leftmost position of the latter to the rightmost position of the former, thus accounting for a redundant path, even though such a path may take longer to get to destination.

That behaviour may be expressed in an algebraic manner by means of a process algebra called Algebra of Communicating Processes (ACP), which is an abstract algebra allowing to express a system as sets of equations expressing how a system works, thus leaving aside its real nature [18]. This way, a model may be built up in order to further extract its external behaviour, which eventually may be compared to that of the real system, leading to the verification of the model if both behaviours run the same string of actions and possess the same branching structure, implying that for every input in the model, the desired output may be obtained, matching that of the real system [19].

In this context, expression (4) denotes the model specification determined in ACP, whereas expression (5) presents external behaviour of the model, whilst expression (6) exposes that of the real system. It is clear that both external behaviours present recursive variables multiplied by the same factors, hence the aforementioned conditions are met, thus the model gets verified.

$$
V_i = \sum_{i=0}^{k^n-1} \left(\sum_{p=0}^{M+2k-1} \left(r_{V_i,p}(d) \cdot \left(s_{V_i,b|M}(d) \triangleleft i = \left\lfloor \frac{b}{M} \right\rfloor \triangleright \left(\sim flag \cdot \sim top \cdot \right. \right. \right. \right.
$$

$$
\sum_{temp=0}^{n-1} \left(\sum_{j=0}^{temp} \left(\left(flag{+}{+} \triangleleft \left\lfloor \frac{i}{k^{temp-j}} \right\rfloor_{|k} \neq \left\lfloor \frac{\lfloor b/M \rfloor}{k^{n-1-j}} \right\rfloor_{|k} \triangleright top{+}{+} \right) \right. \right.
$$

$$
\left. \left. \left. \left. \left. \triangleleft flag = 0 \triangleright \emptyset \right) \right) \cdot s_{V_i,M+\left\lfloor \frac{\lfloor b/M \rfloor}{k^{n-1-top}} \right\rfloor_{|k}}(d) \right) \right) \right) \right) \cdot V_i \quad (4)
$$

$$
\left\| \right.^{k^n-1}_{i=0} \tau_I \left(\partial_H (V_i) \right) = r_{\{V_{\lfloor \frac{a}{M} \rfloor}\},a_{|M}}(d) \cdot s_{\{V_{\lfloor \frac{b}{M} \rfloor}\},b_{|M}}(d) \cdot \tau_I \left(\partial_H (V_i) \right) \quad (5)
$$

$$
\tau_I(X) = r_{\{V_i\},p}(d) \cdot s_{\{V_i\},p}(d) \cdot \tau_I(X) \quad (6)
$$

6   De Bruijn Torus

Extending the concept of de Bruijn sequences from 1 dimension up to 2 dimensions, it is possible to obtain de Bruijn tori, where all possible matrix-like patterns in a certain alphabet are contained exactly once in a toroidal array, such as rows and columns located on the edges may be wrapped around to those situated on the opposite edge, thus representing a sort of a periodic mapping.

Such a de Bruijn torus may have dimensions $r \times r'$, whereas the patterns in an alphabet k to be uniquely found therein may have dimensions $m \times m'$, where initially r and r' may be different, as well as m and m', standing for rectangular matrices, whose nomenclature is $(r, r'; m, m')_k$, where expression (7) holds.

$$r \times r' = k^{m \times m'} \tag{7}$$

It is to be considered that it is possible to construct square de Bruijn tori (where $r = r'$) for square patterns ($m = m'$) of even size, no matter the alphabet being used, or otherwise, if k is a perfect square, where any of both cases results in $(r, r; m, m)_k$.

The shortest square de Bruijn torus is $(4, 4; 2, 2)_2$, where all 16 available submatrices with dimensions 2×2 may be found precisely once within the torus with dimensions 4×4, where the only available configurations are that of a Brigid's cross along with its transposed matrix, considering that rotations over rows and columns make part of the same equivalence class under rotation. In this context, in order for the de Bruijn torus to be a square, it is to be noted that other k-alphabets may be used if the equality in expression (7) holds, along with $r = r'$ and $m = m'$.

The aforementioned expression works with any k-alphabet [20], thus leading to a square de Bruijn torus by meeting $(k^{m^2/2}, k^{m^2/2}; m, m)_k$. For instance, $(256, 256; 4, 4)_2$ and $(2^{18}, 2^{18}; 6, 6)_2$ may be obtained for a binary alphabet, whereas $(9, 9; 2, 2)_3$ and $(16, 16; 2, 2)_4$ may do it for other alphabets. Additionally, squared 2×2 patterns get that expression simplified to $(k^m, k^m; m, m)_k$ [21].

Furthermore, there are other particular cases, such as $(8, 8; 3, 2)_2$, where the torus is square but the pattern is rectangular. Additionally, there are other cases where a rectangular torus is achieved, even though the pattern is squared, such as $(16, 32; 3, 3)_2$. Moreover, some other cases allow for a rectangular torus and a rectangular pattern, such as $(8, 32; 2, 4)_2$. Anyway, all those cases meet the aforesaid expression, although there may be other representations meeting different expressions, or even other kinds of shapes, like an L-shaped pattern [22].

In any case, a periodic mapping of the whole toroidal array is achieved, obtaining a structure of a de Bruijn torus, similar to that depicted in Fig. 7. It may be seen that each node has a link to its predecessor and another one to its successor in both rows and columns, accounting up to 4 neighbour nodes, reminding that all the elements located in each row and each column within a pattern ought to be considered all together as a single block.

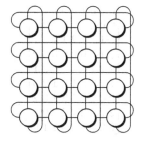

Fig. 7. Two-dimensional toroidal array (4×4 nodes).

Regarding the mapping of the diverse patterns in a square de Bruijn torus, Fig. 8 exhibits an instance of $(4, 4; 2, 2)_2$ along with the mapping of the 16 unique 2×2 binary patterns by moving along the corresponding rows and columns. It is to be noted that the top left element in each pattern may be seen as the handle of it, as its position within the torus, given as rows and columns apart from the top left item of that torus, matches the location of the pattern within the mapping.

$$\begin{bmatrix} 0 & 1 & 0 & 0 \\ 0 & 1 & 1 & 1 \\ 1 & 1 & 1 & 0 \\ 0 & 0 & 1 & 0 \end{bmatrix} \xrightarrow{\text{map}} \begin{bmatrix} \begin{bmatrix} 0 & 1 \\ 0 & 1 \end{bmatrix} & \begin{bmatrix} 1 & 0 \\ 1 & 1 \end{bmatrix} & \begin{bmatrix} 0 & 0 \\ 1 & 1 \end{bmatrix} & \begin{bmatrix} 0 & 0 \\ 1 & 0 \end{bmatrix} \\ \begin{bmatrix} 0 & 1 \\ 1 & 1 \end{bmatrix} & \begin{bmatrix} 1 & 1 \\ 1 & 1 \end{bmatrix} & \begin{bmatrix} 1 & 1 \\ 1 & 0 \end{bmatrix} & \begin{bmatrix} 1 & 0 \\ 0 & 1 \end{bmatrix} \\ \begin{bmatrix} 1 & 1 \\ 0 & 0 \end{bmatrix} & \begin{bmatrix} 1 & 1 \\ 0 & 1 \end{bmatrix} & \begin{bmatrix} 1 & 0 \\ 1 & 0 \end{bmatrix} & \begin{bmatrix} 0 & 1 \\ 0 & 0 \end{bmatrix} \\ \begin{bmatrix} 0 & 0 \\ 0 & 1 \end{bmatrix} & \begin{bmatrix} 0 & 1 \\ 1 & 0 \end{bmatrix} & \begin{bmatrix} 1 & 0 \\ 0 & 0 \end{bmatrix} & \begin{bmatrix} 0 & 0 \\ 0 & 0 \end{bmatrix} \end{bmatrix}$$

Fig. 8. De Bruijn torus $(4, 4; 2, 2)_2$: squared patterns.

As a side note, Fig. 9 exhibits an instance of $(9, 9; 2, 2)_3$, which may account for a total amount of 81 unique 2×2 ternary patterns along the rows and columns within the toroidal array.

$$\begin{bmatrix} 0\,0\,0\,1\,0\,0\,0\,1\,0 \\ 0\,0\,1\,2\,2\,1\,2\,2\,1 \\ 1\,1\,1\,1\,2\,1\,2\,1\,1 \\ 1\,1\,2\,0\,0\,2\,0\,0\,2 \\ 2\,2\,1\,2\,0\,1\,0\,2\,1 \\ 1\,1\,0\,2\,1\,0\,1\,2\,0 \\ 0\,0\,2\,0\,1\,2\,1\,0\,2 \\ 2\,2\,2\,0\,2\,2\,2\,0\,2 \\ 2\,2\,0\,1\,1\,0\,1\,1\,0 \end{bmatrix}$$

Fig. 9. De Bruijn torus $(9, 9; 2, 2)_3$: squared patterns.

7   De Bruijn Hypertorus

Extending the concept of de Bruijn tori from 2 dimensions up to higher dimensions, it is possible to attain de Bruijn hypertori, or hyper-toroidal arrays, even making possible the infinite de Bruijn hypertori.

In this sense, considering $\bar{r} = (r_1 \cdots r_N)$ and $\bar{m} = (m_1 \cdots m_N)$, where $r_i > m_i$ for $1 \leq i \leq N$ and expression (8) holds, which is an extension of expression (7). This way, an N-dimensional hypertoroidal k-ary block may be referred to as $(\bar{r}; \bar{m})_k^N$ de Bruijn hypertorus if it has dimensions $r_1 \cdots r_N$ and every k-ary $m_1 \cdots m_N$ block appears precisely once along the N-dimensional hypertorus [23].

$$\prod_{i=1}^{N} r_i = k^{\prod_{i=1}^{N} m_i} \tag{8}$$

Hence, for any dimension N, it may mean that all possible hypermatrix-like patterns in a certain k-alphabet are contained exactly once in a hypertoroidal array of dimension N, hence representing a sort of a periodic hypermapping.

Furthermore, it is to be considered that it is possible to construct N-hypercube de Bruijn hypertori $(r_1 = \cdots = r_N = r)$ for N-hypercube patterns $(m_1 = \cdots = m_N = m)$, where the conditions proposed for $N = 2$ may be extended herein, resulting in $r_1 = \cdots = r_N = k^{m^N/N}$, where each node has up to $2N$ neighbours.

Focusing on the case of 3 dimensions, the smallest cubic de Bruijn hypertorus with a cubic pattern is $(256, 256, 256; 2, 2, 2)_8$, also named as $(256; 2)_8^3$ [24], such that $(k^{m^3/3}, k^{m^3/3}, k^{m^3/3}; m, m, m)_k$. Furthermore, the smallest rectangular de Bruijn hypertorus is the $16 \times 4 \times 4$ hypertoroidal array, which is filled in with $2 \times 2 \times 2$ cubic binary patterns called b-cubes, which may be named as $(16, 4, 4; 2, 2, 2)_2$ [25], whilst $(81, 9, 9; 2, 2, 2)_3$ may also exist. It is to be reminded that each of those cubic patterns appears just once, such that it is achieved a periodic mapping of the whole three-dimensional de Bruijn hypertoroidal array.

Figure 10 shows a b-cube, where every node has a link to its predecessor and its successor in each dimension. Furthermore, each b-cube has a handle, defined as the front, top, left element, which may help identify each instance of pattern.

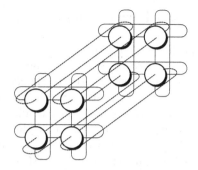

Fig. 10. Three-dimensional toroidal array ($2 \times 2 \times 2$ nodes).

8 k-Ary n-Cubes

A compelling alternative to de Bruijn designs when dealing with overlay networks may be carried out by k-ary N-cubes. These may be seen as hypertoroidal arrays in N dimensions, where up to k nodes belong to each particular dimension [26]. Furthermore, it is to be noted that, when $k = 2$, these designs are just the same as those of the N-hypercube, whereas if $k \geq 3$, a ring may be considered if $N = 1$, a torus if $N = 2$ and a k-ary N-cube (or hypertorus) if $N \geq 3$.

Regarding the k^n nodes in each of those shapes, they may be named by a k-base N-bit identifier d for each dimension $(d_{N-1} \cdots d_i \cdots d_0)$ and they may be connected to all nodes whose identifiers differ in just $d_i \pm 1 \, mod \, k$, accounting for as many neighbours as $2N$, with a pair in each dimension i [27].

In some way, k-ary n-cubes may be seen as a sort of regular undirected graphs, making for redundant paths from a given source to a given destination, which may provide alternative solutions to those furnished by de Bruijn shapes in terms of traffic forwarding and location services, as designs are pretty similar. In fact, a 4-ary 2-cube matches Fig. 8 whereas a 2-ary 3-cube does Fig. 10.

The concept of *handle* may be imported herein, resulting in that being the node $(0 \cdots 0)$, whilst other nodes are named according to how far apart in each particular dimension are from the handle, which acts herein as a reference point, as depicted in Fig. 11 for a 4-ary 2-cube and a 2-ary 3-cube.

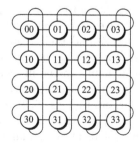 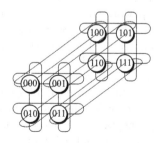

Fig. 11. 4-ary 2-cube toroidal array and 2-ary 3-cube hypertoroidal array.

9 k-Ary n-Cube as a Forwarding Scheme

It is to be noted that both de Bruijn shapes and k-ary n-cubes may be seen as hypertoroidal arrays as their corresponding shapes match, hence, they both may share a similar forwarding scheme. Actually, the algebraic expression given for the former may be applied in the latter with slight modifications, such as in the latter all uplinks may carry bidirectional traffic, and in case of a node failure, that may be substituted with any other neighbouring node in order to force a redundant path. As per the port layout for a generic node, the first M links are downlinks going for localhosts (from 0 to $M-1$) and the last $2N$ links ($M \cdots M + 2N - 1$) are incoming and outgoing links, where M and $M + 1$ are tied to dimension 1, the former heading for its predecessor and the latter for its successor, and the rest of pairs work similarly for the other dimensions.

Regarding the sequence of events when forwarding data, the difference with the previous case comes up when $i \neq \lfloor b/M \rfloor$, where the mismatching symbols may be spotted and traffic may be sent over to them. In that case, each dimension of i is check out and modified in order to align it with its counterpart in $\lfloor b/M \rfloor$.

In this context, expression (9) describes the model specification in ACP, whereas the verification goes just the same way as shown in the de Bruijn case.

$$
V_i = \sum_{i=0}^{k^N-1} \left(\sum_{p=0}^{M+2N-1} \left(r_{V_i,p}(d) \cdot \left(s_{V_i, b_{|M}}(d) \vartriangleleft i = \left\lfloor \frac{b}{M} \right\rfloor \vartriangleright \right. \right. \right.
$$

$$
\left(\left(s_{V_i,M}(d) \vartriangleleft (\frac{-N}{2} \leq (\left\lfloor \frac{b}{M} \right\rfloor_{|N} - i_{|N}) < 0) \ \text{OR} \ ((\left\lfloor \frac{b}{M} \right\rfloor_{|N} - i_{|N}) > \frac{N}{2}) \vartriangleright s_{V_i,M+1}(d) \right) \cdot
$$

$$
\left(\sum_{t=1}^{N-1} \left(s_{V_i,M+2t}(d) \vartriangleleft ((\frac{-N}{2})^t \leq (\left\lfloor \frac{\lfloor b/M \rfloor}{N^t} \right\rfloor - \left\lfloor \frac{i}{N^t} \right\rfloor) < 0) \ \text{OR} \right. \right.
$$

$$
\left. \left. \left. \left. \left. ((\left\lfloor \frac{\lfloor b/M \rfloor}{N^t} \right\rfloor - \left\lfloor \frac{i}{N^t} \right\rfloor) > (\frac{N}{2})^t) \vartriangleright s_{V_i,M+2t+1}(d) \right) \right) \right) \right) \right) \cdot V_i
$$

$$
(9)
$$

10 De Bruijn Shapes as Location Schemes

If a de Bruijn shape is used for position encoding, then de Bruijn decoding needs to be performed by means of an efficient algorithm to do so, being diverse strategies to achieve so [28]. Besides, several instances have been quoted herein so as to appreciate their potential uses locationwise and their relatively easy implementations. Besides, k-ary n-cubes may also make it as they share shapes.

Regarding $1D$ location, it may be performed thanks to the properties of a de Bruijn sequence, as a particular string only appears once in the whole series, no matter if the deployment is rectilinear, circular, or otherwise. This may allow for the implementation of beacons along the trajectory, those being composed of just one sort of signal being on or off (binary alphabet), or as many types as needed (k-ary alphabet). Furthermore, a rolling drum may work as a compass.

With respect to $2D$ location, it may be carried out by means of the properties of a de Bruijn torus, as a particular pattern just shows up once in the whole matrix. It may also be deployed through beacons acting as the desired alphabet.

With regard to $3D$ location, it may be undertaken due to the properties of a de Bruijn hypertorus, as a given cubic pattern just takes place once in the whole volume structure. As in the other cases, it is done by using the proper beacons.

11 Conclusions

In this paper, de Bruijn shapes have been studied, and further compared to k-ary n-cubes, as both topologies are quite popular when dealing with overlay networks, thus decoupling network services out of their real infrastructure.

Focusing on fog computing, those topologies may offer an easy way to forward traffic in ecosystems with many network devices by means of taking advantage of the regular properties of those designs as hypertoroidal arrays of a determined dimension. This way, receiving and sending packets may be easily done by applying basic arithmetic operations.

On the other hand, both topologies offer location services in one, two and three dimensions just by using a determined series of beacons, hence, obtaining the location of an item within the fog by means of spotting singular patterns. In this sense, several examples have been presented so as to better understand how to work out the location of an item within a fog domain.

References

1. Prabhu, C.S.R.: Overview - fog computing and Internet-of-Things (IOT). EAI Endors. Trans. Cloud Syst. **3**(10), 1–23 (2017). https://doi.org/10.4108/eai.20-12-2017.154378
2. Anwar, M., et al.: Fog computing: an overview of big IoT data analytics. Wirel. Commun. Mobile Comput. **2018**, 7157192 (2018). https://doi.org/10.1155/2018/7157192
3. Bittencourt, L.F., Lopes, M.M., Petri, I., Rana, O.F.: Towards virtual machine migration in fog computing. In: Proceedings of 10th International Conference on P2P, Parallel, Grid, Cloud and Internet Computing (3PGCIC), pp. 1–8 (2015). https://doi.org/10.1109/3PGCIC.2015.85
4. Osanaiye, O., Chen, S., Yan, Z., Lu, R., Choo, K.R., Dlodlo, M.: From cloud to fog computing: a review and a conceptual live VM migration framework. IEEE Access **5**, 8284–8300 (2017). https://doi.org/10.1109/ACCESS.2017.2692960
5. Mahmud, R., Ramamohanarao, K., Buyya, R.: Application management in fog computing environments: a taxonomy, review and future directions. ACM Comput. Surv. **53**(4), pp. 1–43 (2020). https://doi.org/10.1145/3403955
6. Darlagiannis, V., Mauthe, A., Heckmann, O., Liebau, N., Steinmetz, R.: On routing in a two-tier overlay network based on de Bruijn digraphs. In: IEEE/IFIP Network Operations and Management Symposium (NOMS), pp. 186–197 (2006). NOMS.2006.1687550

7. Richa, A., Scheideler, C., Stevens, P.: Self-stabilizing De Bruijn networks. In: Défago, X., Petit, F., Villain, V. (eds.) SSS 2011. LNCS, vol. 6976, pp. 416–430. Springer, Heidelberg (2011). https://doi.org/10.1007/978-3-642-24550-3_31
8. Galán-Jiménez, J., Gazo-Cervero, A., Overview and challenges of overlay networks: a survey. Int. J. Comput. Sci. Eng. Surv. (IJCSES) **2** (1), 19–37 (2011). https://doi.org/10.5121/ijcses.2011.2102
9. West, R., Fry, G., Wong, G.: Comparison of k-ary n-cube and de Bruijn Overlays in QoS-constrained Multicast Applications. In: Proceedings of the International Conference of Parallel and Distributed Processing Techniques and Applications (PDPTA), pp. 1–11. (2005). https://hdl.handle.net/2144/1829
10. de Bruijn, N.G.: A combinatorial problem. Proc. Koninklijke Nederlandse Akademie V. Wetenschappen **49**, 758–764 (1946). https://www.dwc.knaw.nl/DL/publications/PU00018235.pdf
11. Chee Y.M., Etzion, T., Kiah, H.M., Vu, V.K., Yaakobi, E.: Constrained de Bruijn codes and their applications. In: IEEE International Symposium on Information Theory (ISIT), pp. 2369–2373 (2019). https://doi.org/10.1109/ISIT.2019.8849237
12. Almansi, E., Becher, V.: Completely uniformly distributed sequences based on de Bruijn sequences. Math. Comput. **89**, 2537–2551 (2020). https://doi.org/10.1090/mcom/3534
13. Kociumaka, T., Radoszewski, J., Rytter, W.: Efficient ranking of Lyndon words and decoding lexicographically minimal de Bruijn sequence. In: SIAM J. Discrete Math. **30**(4), 2027–2046 (2016). https://doi.org/10.1137/15M1043248
14. Lytle, M.: Edge-disjoint hamiltonian cycles in De Bruijn graphs. Thesis degree of Honors Baccalaureate of Science in Mathematics, Oregon State University, pp. 1–19 (2013). https://ir.library.oregonstate.edu/downloads/8049g733b?locale=en
15. Wong, C.H.: Novel universal cycle constructions for a variety of combinatorial objects. Ph.D. Thesis in Computer Science, University of Guelph, pp. 1–117 (2015). https://atrium.lib.uoguelph.ca/xmlui/handle/10214/8812
16. Sawada, J., Williams, A, Wong D.: A simple shift rule for k-ary de Bruijn sequences. Discrete Math. **340**(3), 524–531 (2017). https://doi.org/10.1016/j.disc.2016.09.008
17. Darlagiannis, V., et al.: On routing in a two-tier overlay network based on de Bruijn digraphs. In: IEEE/IFIP Network Operations and Management Symposium, pp. 186–197 (2006). https://doi.org/10.1109/NOMS.2006.1687550
18. Groote, J.F., Mousavi, M.R.: Modelling and Analysis of Communicating Systems, pp. 1–392. MIT Press, London (2014). https://dl.acm.org/citation.cfm?id=2628007
19. Fokkink, W.: Introduction to Process Algebra, pp. 11–74. Springer, Berlin (2000). https://doi.org/10.1007/978-3-662-04293-9
20. Horan, V., Stevens, B.: Locating patterns in the De Bruijn Torus. Discrete Math. **339**(4), 1274–1292 (2016). https://doi.org/10.1016/j.disc.2015.11.015
21. Hurlbert, G., Isaak, G.: On the De Bruijn torus problem. J. Combin. Theory Ser. A **64**(1), 50–62 (1993). https://doi.org/10.1016/0097-3165(93)90087-O
22. Pudwell, L., Rockey, R.: de Bruijn arrays for L-fillings. Math. Mag. **87**(1), 57–60 (2014). https://doi.org/10.4169/math.mag.87.1.57
23. Kapinya, J.B.: Evolutionary computing solutions for the de Bruijn torus problem. Master's thesis in Computer Science, University of Vrije, pp. 1–33 (2004). http://compalg.inf.elte.hu/~tony/Kutatas/PerfectArrays/Kapinyathesis.pdf
24. Horváth, M., Iványi, A.: Growing perfect cubes. Discrete Math. **308**(19), 4378–4388 (2008). https://doi.org/10.1016/j.disc.2007.08.031

25. Casteels, K., Tinker, T.: De Bruijn sequences of higher dimension. Master's thesis of Ted Tinker, directed by Karel Casteels, University of California, pp. 1–41 (2018). http://web.math.ucsb.edu/~casteels/TedTinkerSenior_Thesis.pdf
26. Mao, W., Nicol, D.M.: On k-ary n-cubes: theory and applications. Discrete Appl. Math. **129**(1), 171.193 (2003). https://doi.org/10.1016/S0166-218X(02)00238-X
27. Shih, Y.K., Kao, S.S.: One-to-one disjoint path covers on k-ary n-cubes. Theor. Comput. Sci. **412**, 4513–4530 (2011). https://doi.org/10.1016/j.tcs.2011.04.035
28. Boutin, D., Horan, V., Pelto, M.: Identifying codes on directed de Bruijn graphs. Discrete Appl. Math. **262**, 29–41 (2019). https://doi.org/10.1016/j.dam.2019.02.005

E-services

Assistive e-Learning Software Modules to Aid Education Process of Students with Visual and Hearing Impairment: A Case Study in North Macedonia

Kostadin Mishev[1]([✉]), Aleksandra Karovska Ristovska[2],
Olivera Rashikj-Canevska[2], and Monika Simjanoska[1]

[1] Faculty of Computer Science and Engineering, Ss. Cyril and Methodius University,
Skopje, North Macedonia
{kostadin.mishev,monika.simjanoska}@finki.ukim.mk
[2] Faculty of Philosophy, Ss. Cyril and Methodius University,
Skopje, North Macedonia
{aleksandrak,oliverarasic}@fzf.ukim.edu.mk

Abstract. This paper presents a technology4good initiative that integrates multiple breakthrough software modules with the aim to build a generic framework to aid the educational process for students with disabilities, such as, hearing and vision impairments, as well as various types of dyslexia. The purpose of the study is to apply various distinct researches among which the highlight is on the text-to-speech engine for the first time developed to support Macedonian language. Additionally, the framework integrates Macedonian sign language to guide the hearing impaired students through the contents of the educational framework. Also, there is a specially developed font (typeface) and color environment for students with specific reading difficulties (dyslexia). The methods that support the educational framework are developed by mix of social, special education and computer science experts. The methodology for developing the text-to-speech engine relies on the newest and most efficient principles in Machine Learning for Natural Language Processing - a Deep Learning approach. The framework has been tested on target group of students and the satisfaction has been measured by using the standard Likert scale.

Keywords: Technology4good · Educative framework ·
Text-to-speech · Deep learning · Natural language processing ·
Assistive technologies

1 Introduction

A rapid increase in the use of personal computers occured in 1970s s and 1980s.s. This coincides with the same time frame, when several societal movements began towards providing equality and equity for disabled persons [16]. Following the

© Springer Nature Switzerland AG 2022
L. Antovski and G. Armenski (Eds.): ICT Innovations 2021, CCIS 1521, pp. 145–159, 2022.
https://doi.org/10.1007/978-3-031-04206-5_11

movement towards inclusion, students with different types of disabilities have been included in mainstream schools. General teachers from mainstream classrooms seek ways to give additional support for students which have difficulties with comprehension of the study material and decoding of words, particularly when they access curricular materials from different content areas. Advancement of the personal computer has had a consequential influence on the provision of services for persons with disabilities. There are inconsistent regulations related to Web Accessibility in different countries due to the divergent IT and disability policies [6]. The Convention on the Rights of Persons with Disabilities, signed by more than 150 countries, in Article 9 implicates the commitment to "promote access for persons with disabilities to new information and communication technologies and systems, including the Internet", and "promote the design, development, production and distribution of accessible information and communications technologies and systems at an early stage, so that these technologies and systems become accessible at minimum cost" [3]. Today, there is a wide range of accessibility options and technologies for persons with different type and level of disability such as mechanical joysticks, vicinity sensors, adapted track-balls, voice recognition software, wearable sensors for electroencelography and electromyography. Related to education, there are cost and time-effective technologies which enable students to access the mainstream curricula. On the other hand, if an assistive technology is available, this does not necessarily mean that it strengthens learning, or, is adequate for each disabled student. Selecting the best-fitting educational technology for the classroom is crucial to the success of the students who are using it [31]. In addition to the large number of different types of disability, reading as a multiplex cognitive task is one of the most common reason for elevation of dropout of students from schools throughout the world [2,37]. Many students enrolled in high schools have difficulties with fluent reading, particularly comprehension. This is largely due to the slow reading rate and inconsistent decoding as well as the large number of mistakes they make during the reading process. Additionally, many of the students show a deficit in content-related vocabulary. The gap in vocabulary only grows throughout the years. Students with high achievements have shown to know four times as many words as students with low performances. Students with poor reading skills or no access to tools which would improve their reading, are more likely to become drop-outs by high school may, because they are unable to keep the same pace with the coursework as their peers [37]. So, students with disabilities who have difficulties reading texts from their grade level, or students with specific reading and writing difficulties as well as blind students, can improve their comprehension in reading by using assistive technology which reads texts aloud [7].

In the past ten or so years, speedy innovations in text-to-speech (TTS) technology have given new and low-cost ways to aid disabled students read printed or digital texts with no audio alternatives. TTS technology enables students to effectively hear any text which is read aloud with a synthesized voice expression [35]. TTS technology is an example of assisted technology (AT) that has become a frequent tool used by students that struggle with reading in mainstream schools, and is has been universally accepted as a form of individualization and accommodation for disabled students. It is important to look for accuracy in the conversion of text in reading materials, with tools like Optical Character Recognition (OCR), and the quality of the TTS voices. When the words are being presented auditorily, the student focuses on the word meaning rather then taking up all their brainpower in trying to vocalize words. The TTS software explores the text by the use of a system of phonemes, phonics and other rules for identification of words, and then the software reads aloud any text via voice synthesis. Users are given a synchronized text presentation in two different modalities - both visual (text on screen) and auditory (hearing words spoken aloud) [41]. Numerous commercial TTS software packages are available in various languages, however, prior to our latest research on the subject [26] which details are presented in the later sections, there was no appropriate TTS software able to reproduce the content of the screen in Macedonian language with a human-like voice, not the existing classic robot voice.

In this paper the focus is on the AT research realization in form of software modules that are integrated in e-Learning platform with the purpose to ease and make all the contents accessible to students with disabilities. The rest of the paper is organized as follows. The efficiency of the TTS systems in education improvement of students with disabilities is discussed in Sect. 2. The dyslexia, sign language and Macedonian TTS synthesizer research is provided in Sect. 3. Section 4 presents the architecture of the modules integrated in the e-Learning platform. The same modules are integrated by a developed mobile application which is described in Sect. 5. The satisfaction and efficiency is measured by specially chosen group of students and the results are presented in Sect. 6. Final Sect. 7 presents the conclusions and directions for future work.

2   Related Work

The effect of TTS on improving comprehension in reading and education of students with disabilities has been widely studied in the last 30 years [27]. Speech synthesis as a category of software that converts text to artificial speech has a wide range of components that can aid in the reading process. Text-to-speech software and similar technologies for reading text aloud are largely implemented with the purpose to assist skills for reading and comprehension in students with various types of disabilities or specific learning difficulties [41]. Reading and reading comprehension is mostly difficult for students that have specific reading disabilities [20], and the most present-day theories claim the most prevalent cause and a predictor for reading difficulties is the problems they encounter while decoding written chunks of text. The lack of decoding skills has a negative effect on reading and comprehension and leads to a decrease in accuracy and reading speed. When course-work is presented orally, supplementary to the traditional manner of presentation, it withdraws the need for decoding of chunks of text and it helps students with specific reading difficulties with their reading comprehension. The results of the Wood's study [41] suggest that text-to-speech software may assist students with their reading comprehension. Melisa Oberembt in [28] has made a review analysis of the twenty-eight studies (peer-reviewed) which were published between 2002 and 2019, elaborating the topic of the effects of the use of text-to-speech for students with disabilities. The research indicated that rates of reading increased, while there were mixed results regarding reading comprehension. Writing skills did not notably improve during the use of this technology. The research also established the social soundness of TTS. Hebert and Murdock [15] did a research related to vocabulary acquisition in students with specific learning disabilities within three dissimilar situations: absence of speech, digitized speech and synthesized speech. Three students with learning disabilities from sixth grade were introduced to 25 words per day for 6 days by using one of the treatments. The results showed that when students used digitized speech they had a better vocabulary acquisition opposite to reading text while not being aided by a digitized voice. Dawson, Venn, and Gunter [13] did a comparison of fluency in reading in students with behavioral difficulties. Three groups were included: students that weren't using any reading models as assistance, students that were modeling their reading according to the teachers' model of reading, and students using TTS technology. This research led to similar results in the three groups: the highest rate of word read accurately was achieved by using the teachers' model, but the highest number of precisely read words while not using the teacher as a reading model was achieved with the use of TTS. Monica C. Silió & Barbetta [34] in their research related to Hispanic boys with learning difficulties from the fifth grade confirmed that word prediction in a combination with a TTS model improved their writing skills and the composition of narratives. [29] Parr in his research assess the attitude of the students toward TTS, determining the future of the TTS technology. In his conclusions he wrote that when students use TTS as a part of a more comprehensive wholesome approach, they integrate this experience and embed it into their meta-cognitive strategies. This improves their efficacy and self-advocacy and gives them means for collaboration with their peers. TTS is a software that for some will be a marginalized tool, for others a choice in the learning process, but for some it will be lifelong instrument.

3 Methods

This section explains the methods with which the software modules were developed and integrated in the e-learning platform. All the methods such as the Macedonian TTS system, the Macedonian sign language representation, and the Macedonian dyslexic font are originally developed for the e-learning platform needs and are considered pioneers for Macedonian language.

3.1 Macedonian Language TTS Synthesizer

In order to meet the needs of our software module, we needed to develop Macedonian TTS system from scratch and we relied on completely different methodology than any other trial before. Thus, we used the leverage of the Deep Learning approach and built a synthesizer that sounds natural and human-like, named

MAKEDONKA [26]. Among the variety of Deep Learning-based methods, Deep Voice 3 was chosen to be most appropriate solution for the creation of Macedonian language TTS system, since it is able to synthesize more than 10M sentences per day [32], and thus is convenient for each day usage by the students who need its service in the process of learning. This ability is very important considering the purpose of its development - to help the students with visual impairment. This help means that sometimes it will be used only for navigation via the education platform, but also it could be used for "reading" entire books uploaded by the professors. To achieve single-speaker synthesis, approximately 20 h of Macedonian high-quality speech audio dataset was created. The dataset has been preprocessed by following the example of the golden standard for training English language TTS systems - the LJSpeech dataset [18]. The total training time took 21 days and 16 h. As a results, we achieved a model that produce intelligible human-like speech with no "robotic" components, which was found as very irritating in the previous attempts. The model has been evaluated on seven different sentences with certain specifics, covering variety of cases in the Macedonian language, such as: long/complex words in terms of pronunciation, compound words, comma somewhere in the sentence to check whether the model makes the appropriate pause when synthesizes the speech, sentences ending with fullstop, question mark and exclamation mark to check whether the model is able to change the intonation in the synthesized speech, tongue twisters, and tongue twisters containing words with multiple adjacent consonants. The quality of model was assessed by 15 distinct listeners who gave their subjective opinion on the quality of the synthesized speech. The obtained mean opinion score (MOS) for the ground truth and the selected TTS model, are 4.6234 and 3.9285, correspondingly. Thus, according to the MOS method [33,39], the obtained value indicates a good quality audio, with no additional effort to understand the words, distinguishable sounds, not annoying pronunciation, preferred speed, and pleasant voice.

Creation of this model [26] is the first successful attempt for Macedonian language TTS synthesizer creation in the last 30 years. We are the first who built it and publish the model available to use as a module in any kind of software that needs its service. Further sections describe its usage as a module in education software to aid the learning process for students with visual impairment.

3.2   Macedonian Sign Language

Sign languages of deaf communities all around the globe are fully developed human languages with a full expression. Nevertheless, signs were once viewed as a system of graphic gestures without linguistics embedded in their structure [14]. In the past, sign languages have been disputed in linguistic research and haven't been defined as real languages. This was due to the differences in sentence production in sign and spoken languages. Like all spoken languages, sign languages have grammar and linguistic structure. Sign languages do not follow the same grammatical patterns as spoken languages and there is a need for a substantially different conception of grammar [23]. This makes translation of signed languages to spoken languages and vice versa a very complex issue, because unlike translation in spoken languages where there is a mapping word for word, here we have the use of a completely different modality [38]. Sign languages utilize a visual modality for their expression. Spoken languages are naturally sequential, and there is a word order that needs to be considered.

In the 1970s and 1980s, which were the onset for linguistic research for sign languages, it was discovered that they also have morphology which is vastly complex. Research data shows that the morphology in signed languages is simultaneous and morphemes are superimposed unlike the spoken languages where words are forming a string structure [8]. If we are trying to compare signed to spoken languages, the easist way to do this is through comparison of the syntax. There are prevalent similarities in the typology of spoken vs sign languages in the domain of syntax [36]. Neidle et al. [1] point out that sign languages not only have a very liberal word order, but also, like some spoken languages, they do not have a hierarchic phraseological structure but use other principles (fluency articulation, economic saving by using articulation). Cecchetto and his colleagues [10] propose that different sign languages have different syntax rules which derives from the non-manual markers which are used as grammatical markers.

Modern research shows that non-manual markers in sign languages can be compared to the prosody in spoken languages [12]. Movements of the brows or specific head movements are used for yes or no questions, or different types of content questions in almost all sign languages. These markers in sign language and analogue to intonation makers in most spoken languages. Hence, the translation of spoken languages to sign languages is very complex, and it cannot be achieved simply by mapping a sign to a certain word. Sign languages use articulators (the hands) to convey meaning. These so called articulation channels use movement of the hands, shape of the hands, and movement of the upper body as well as non-manual markings (eye gaze, expressions of the face, mouthing words, posture of the body).

Macedonian Sign Language, as a member of the large family of sign languages is a natural language [4] used by the Deaf community in Republic of North Macedonia, or, by approximately 6000 deaf individuals. Commercial applications for many sign languages largely focus on gestures not signs, and they simply map signs or gestures to spoken words [38]. This is due to the misapprehension that the deaf population is comfortable reading spoken language and does not need translation into sign language. However, there is no guarantee that someone whose first language is, for example, Macedonian Sign Language, is familiar with written Macedonian, as the two are completely separate languages. Using the expertise of deaf community members that understand both the Macedonian sign and written languages, we have developed a module that will ease the work on the learning platform as explained in Sect. 4.

3.3   Macedonian Dyslexic Font

Approximately 5 to 10% of the world population is said to have dyslexia [22]. Persons with dyslexia have difficulties with recognizing printed words, lack fluency in reading, have difficulties pronouncing unknown words, and read very slow. In some European languages that have more regular writing systems (transparent orthographies) than the English language, main indicators for dyslexia are lack of fluency in reading and difficulties with the spelling. Nevertheless, the predictors for reading (and onset of dyslexia) are identical and are related to letter comprehension, awareness of phonemes and skills for rapid naming (RAN) [17]. Dyslexia occurs in all linguistic systems. The prevalence is significantly lower in populations in which the written language is more phonetically consistent, such as Italian, Greek and Czech [30]. In these linguistic populations dyslexia largely manifests as poor fluency in reading. In North Macedonia, dyslexia is a widespread learning disability that impairs the ability to read fluently which in turn translates to a large number of students that have difficulties with their academic achievements.

In the past years, studies have been conducted with the purpose to define the usefulness of specially created typefaces (and subsequently fonts) for persons with dyslexia. In a research leaded by Bachman and Mengheri [9] it was discovered that reading improves in dyslexic readers when changing the typeface from Times New Roman to EasyReading^{TM}. The results show a statistically significant difference in the achievements; the readers that used the EasyReading^{TM} font had a better reading fluency in all given tests (words, non-words and large texts). A research conducted in the US, on the OpenDyslexic font on the other hand showed that there was no advancement in the rate of reading and there wasn't a lower number of mistakes made in the individual groups of student with dyslexia, nor in the group as a whole [40]. A study for the typeface named "Dyslexie", which nowadays in used in many mainstream schools in the Netherlands, showed that enlargement of the font makes the reading easier, regardless of which font or typeface is used [24]. In a study conducted by Masulli et al. [25], it was shown that when students are reading words whose letters are larger, and the space between them is bigger, the number of fixation of the eyes decreases. This would be very helpful for dyslexic readers which are prone to making much more eye fixation than the non-dyslexic readers which in turn impedes their reading speed and accuracy.

Research conducted in the previous years shows that more stable orthographies have bigger use of specific typefaces than opaque orthographies and regardless of the typeface use, the reading fluency of persons with dyslexia benefits from increased spacing and font size. Leaded by this argument, a group of researchers from North Macedonia created the first font for the Cyrillic alphabet for students with dyslexia (and specific reading difficulties) in 2018. The font was named DyslexicFZF, and it was originally built on the Latin font Open Dyslexic. This font has features that can be found is all other fonts for persons with dyslexia. DyslexicFZF is a Sans Serif font, and it was purposefully produced for letters found in the Macedonian alphabet, as well as for numerical and punctuation signs. The font has thickening of the foundation of each letter and there is an enlargement of all the characters. Also the font has increased spacing which leads to decrease of overcrowding within the text. Because of the accentuated difference in the proportion of thinner and thicker lines, the font has a notable distinction [19]. A subsequent study was conducted in North Macedonia, with the purpose to explore the benefit of DyslexicFZF font. The results were in line with the above-mentioned research of the influence of specialized fonts in transparent orthographies. The respondents with dyslexia, read more fluently, made less mistakes and read a larger number of words with the DyslexicFZF font than with the Times New Roman font [5]. The development of the Macedonian DyslexicFZF font was crucial for the successful creation of the dyslexia module which accompanied with palette of colors was integrated into the e-learning platform. Details can be found in the following Sect. 4.

4   Integration with e-Learning Platform

This section provides details of the software modules development as well as their integration into the e-learning platform. The platform is up and running and can be accessed on https://courses.fzf.ukim.edu.mk.

Figure 1 presents the home page of a user after login with his/her credentials. On the right side of the figure, are the three modules integrated into the Moodle e-learning platform: the Macedonian language synthesizer, the dyslexia module and the Macedonian sign language module. Each of them can be explicitly enabled by clicking on the corresponding button. For the dyslexia module, this button is presented in dyslexic font in order to be visible for the users. Once a module it enabled, it remains enabled until the user decides to disable it and return to a default mode.

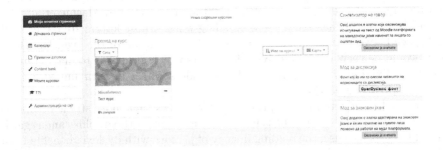

Fig. 1. Home page after login.

4.1   Visual Impairment Module

The visual impairment module refers to two target groups: those that have completely, or, partially lost their sight, and those that have difficulties reading the content of the e-learning platform because of any reason. In the previous Sect. 3 we explained the methodology used to develop MAKEDONKA - the first TTS synthesizer in Macedonian language. This intelligent synthesizer model resides as a service in the Cloud and serves the visual impairment module on request. Those requests are generated any time the mouse reaches a text label on the platform, meaning the page is not completely read as often is the case with other TTS systems implemented in some web pages. Instead, the synthesizer reads only the content at the paragraph pointed by the mouse. Even more, it is also able to describe the content even if it is given in a form of image, by reading the description of the image provided by the moderator which is not visible to the users. This option complies with the Web Content Accessibility Guidelines (WCAG) 2.1 standard [21] which defines the rules for representing the web content to persons with disabilities.

4.2   Hearing Impairment Module

Figure 2 presents the hearing impairment module when activated. When enabled, this module provides Macedonian sign language service to the student. Most of the important features of the platform were captured by an expert who perfectly understands both the Macedonian written and Macedonian sign language. Once the mouse points at some label, the video is started at the lower right corner of the screen, thus the textual content is not hidden from the student, but it is only additionally explained in sign language.

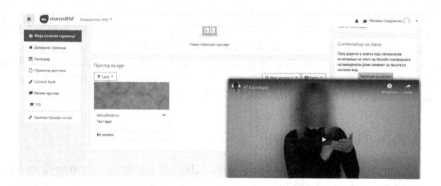

Fig. 2. Macedonian sign language module.

4.3   Dyslexia Module

Enabling the dyslexia module provides an interface as shown in Fig. 3. The DyslexicFZF font developed by Karovska et al. [19] is accompanied with palette of colors that should additionally ease the reading of the students with dyslexia as there are many dyslexia typefaces. Changing the color changes the background of the paragraphs and thus provides more convenient contrast for the dyslexic students. The palette is selected by the experts in the field.

Fig. 3. Macedonian dyslexic font implementation.

5 Mobile Application

In the literature review provided by Chelkowski et al. [11], the importance of the mobile devices for people with disabilities is clearly shown, especially for the visually impaired people. Considering the common keyboard and mouse inadequacy, we have decided to provide the same functionalities of the presented Moodle e-learning platform as custom-based mobile application that differs from the official Moodle application that exists on Google Play Store. The interface of the application is shown on Fig. 4. Starting from the welcome screen at the most left, the module for dyslexia, sign-language, and TTS can be seen. For the user to know the current mode, special icons are inserted next to the active labels at the corresponding mode. All the mouse functionalities are made to be touch-based at this application. The application is active and can be downloaded on Google Play Store by the name $Courses@FZF$ at the link https://play.google. com/store/apps/details?id=mk.edu.ukim.fzf.moodlefzf.

6 Evaluation Results

To evaluate the students' attitude towards the e-learning platform and to confirm that software modules fulfill their needs, we conducted distinct surveys for each target group that contained 13 questions each. The answers describing their attitude and satisfaction were offered according to the Likert scale starting from 1 (Strongly disagree), 2 (Disagree), 3 (Undecided), 4 (Agree) and 5 (Strongly Agree). Tables 1, 2, and 3 present the results in terms of the percentage of students whose answer corresponds to each of the offered answers.

Table 1 presents the satisfaction of the students with visual impairment. As seen from the results, the students strongly agree on the appropriateness of the quality of the module and the overall e-learning platform, referred to as learning management system (LMS) in the surveys. The accessibility of the module, the activation, the settings, the notification, as well as the impact on the communication, are somewhat discussed by the students and it can be seen that there is some room for interface improvements.

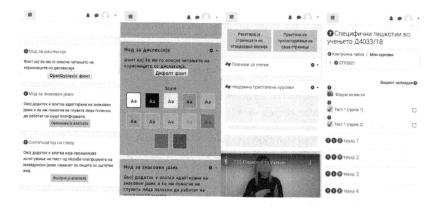

Fig. 4. Mobile application interface.

Table 2 presents the satisfaction results from the group with hearing impairment. Compared to the results from the visually impaired group, in this case it can be seen that this group has no problem with activating the module and similar settings. Instead, what is more difficult to be understood is the format of

Table 1. Students' TTS module evaluation according to Likert's scale

No.	Question	1	2	3	4	5
1	Do you find the accessible Moodle easy to navigate with the help of the TTS module?	0	0	0	50	50
2	Do you find appropriate the activation of the TTS module?	0	0	50	50	0
3	Do you find appropriate the voice of the TTS module?	0	0	0	100	0
4	Do you find appropriate the quality of the TTS module?	0	0	0	0	100
5	Do have easy access to the LMS and its content by using the TTS module?	0	0	0	0	100
6	Is the LMS easily understandable considering the adjustments made for your special needs?	0	0	0	50	50
7	Are you satisfied with the format of the course content being are posted on the LMS?	0	0	0	0	100
8	Are you satisfied with the LMS notification system?	0	0	50	0	50
9	Are you satisfied with the video-conferencing options of the LMS and are they adequate to your special needs?	0	0	0	0	100
10	Is the teaching staff responsive and communicating in a manner adequate to your special needs?	0	0	50	0	50
11	Has your productivity increased after the use of this LMS?	0	0	0	0	100
12	Have your digital skills been improved after the use of the LMS?	0	0	0	0	100
13	Has your communication with your non-disabled peers been improved through the use of this LMS?	0	0	50	50	0

Table 2. Students' Sign language module evaluation according to Likert's scale

No.	Question	1	2	3	4	5
1	Do you find the accessible Moodle easy to navigate with the help of the sign language module?	0	0	0	0	100
2	Do you find easy the activation of the sign language module?	0	0	0	0	100
3	Do you find the position of the videos to be user friendly and easy to follow?	0	0	0	0	100
4	Do you find there are enough videos to cover all the functionalities of the Moodle interface?	0	0	0	50	50
5	Is your access to the LMS and its content easier after implementing sign language module?	0	0	0	0	100
6	Is the LMS easily understandable considering the adjustments made for your special needs?	0	0	0	0	100
7	Are you satisfied with the format of the course content being are posted on the LMS?	0	0	50	50	0
8	Are you satisfied with the LMS notification system?	0	0	0	50	50
9	Are you satisfied with the video-conferencing options of the LMS and are they adequate to your special needs?	0	0	0	0	100
10	Is the teaching staff responsive and communicating in a manner adequate to your special needs?	0	0	50	0	50
11	Has your productivity increased after the use of this LMS?	0	0	0	0	100
12	Have your digital skills been improved after the use of the LMS?	0	0	0	0	100
13	Has your communication with your non-disabled peers been improved through the use of this LMS?	0	0	50	50	0

Table 3. Students' Dyslexic module evaluation according to Likert's scale

No.	Question	1	2	3	4	5
1	Do you find the accessible Moodle easy to navigate with the help of the dyslexic module activation?	0	0	0	0	100
2	Do you find appropriate the colors implemented as a background?	0	0	0	0	100
3	Do you find appropriate the implemented font?	0	0	0	0	100
4	Do you find the module activation button easy to spot?	0	0	0	0	100
5	Has the module made the access to the LMS and its content easier?	0	0	0	50	50
6	Is the LMS easily understandable considering the adjustments made for your special needs?	0	0	0	0	100
7	Are you satisfied with the format of the course content being are posted on the LMS	0	0	50	0	50
8	Are you satisfied with the LMS notification system?	0	0	0	0	100
9	Are you satisfied with the video-conferencing options of the LMS and are they adequate to your special needs?	0	0	0	0	100
10	Is the teaching staff responsive and communicating in a manner adequate to your special needs?	0	0	0	0	100
11	Has your productivity increased after the use of this LMS with dyslexic module implemented?	0	0	0	0	100
12	Have your digital skills been improved after the use of the LMS?	0	0	50	50	0
13	Has your communication with your non-disabled peers been improved through the use of this LMS?	0	0	100	0	0

the course content itself. This is understandable to some extent, since we have prior knowledge that even without visual impairment, this group needs more descriptive approach to represent the contents and this should be taken into account for further improvements of the interface.

The satisfaction of the students with dyslexia is provided in Table 3. As expected, this group has no problems with any of the functionalities of the module, or, the platform in general. However, the format of the content still remains a problem for this group. What is clearly visible for all the groups is that this platform has not intrigued remarkable improvement in the communication with the other non-disabled peers. This might be as a consequence of inactivity of the other student considering the forum options of the platform. This, however, is a problem which is out of scope for this project and should be deeply analysed before appropriate advancements on the interface are made in order to achieve higher satisfaction.

7   Conclusion and Future Work

This paper presents a research that found its applicability in the assistive technologies field and is a real example of technology4good that should be followed by all software engineers. Making the content accessible for everyone should be considered a standard in any society.

The research presents the development and integration of three distinct modules in the popular Moodle e-learning platform. In order to encompass as much target groups, the modules are designed to aid the learning process of students with visual and hearing impairment. At the background of each module there is a comprehensive research including a deep learning approach to train high quality text-to-speech system. Also, experts are consulted for appropriate development of the sign-language and dyslexia modules.

Considering the fact that those target groups are more prone to using mobile phones, we also developed a mobile application that supports the usage of the modules considering the access constraints by the technology, i.e., everything is accessible on touch.

Each module is evaluated by a set of 13 questions reflecting the satisfaction of each target group. The analysis of the results provide some valuable conclusions on the students' perception of the usefulness of the modules and therefore, should be considered as directions for future improvement of the interface of the whole e-learning system in general.

References

1. Neidle, C.J., Kegl, J., Bahan, B., MacLaughlin, D., Lee, R.G.: The Syntax of American Sign Language: Functional Categories and Hierarchical Structure (2000)
2. Kamil, M.L.: Adolescents and Literacy: Reading for the 21st Century (2003)
3. Convention on the rights of persons with disabilities (2008). http://www.un.org/disabilities/default.asp?navid=15&pid=162

4. Comparative analysis of the structure of American and Macedonian Sign Language. Ph.D. thesis (2014). http://hdl.handle.net/20.500.12188/2808
5. DyslexicFZF (2019). http://shorturl.at/PV458
6. Amado-Salvatierra, H., Hernández, R., Hilera, J.: Implementation of accessibility standards in the process of course design in virtual learning environments. Procedia Comput. Sci. **14**, 363–370 (2012)
7. Anderson-Inman, L., Horney, M.A.: Supported eText: assistive technology through text transformations. Read. Res. Q. **42**, 153–160 (2007). https://doi.org/10.1598/RRQ.42.1.8
8. Aronoff, M., Meir, I., Sandler, W.: The paradox of sign language morphology. Language **81**(2), 301–344 (2005). https://doi.org/10.1353/lan.2005.0043
9. Bachmann, C., Mengheri, L.: Dyslexia and fonts: is a specific font useful? Brain Sci. **8**(5), 89 (2018)
10. Cecchetto, C., Geraci, C., Zucchi, S.: Another way to mark syntactic dependencies: the case for right-peripheral specifiers in sign languages. Language **85**, 278–320 (2009)
11. Chelkowski, L., Yan, Z., Asaro-Saddler, K.: The use of mobile devices with students with disabilities: a literature review. Prev. Sch. Fail.: Altern. Educ. Child. Youth **63**(3), 277–295 (2019)
12. Dachkovsky, S., Sandler, W.: Visual intonation in the prosody of a sign language. Lang. Speech **52**(2–3), 287–314 (2009)
13. Dawson, L., Venn, M.L., Gunter, P.L.: The effects of teacher versus computer reading models. Behav. Disord. **25**(2), 105–113 (2000)
14. Goldin-Meadow, S., Brentari, D.: Gesture, sign, and language: the coming of age of sign language and gesture studies. Behav. Brain Sci. **40**, e46 (2017). https://doi.org/10.1017/S0140525X15001247
15. Hebert, B.M., Murdock, J.Y.: Comparing three computer-aided instruction output modes to teach vocabulary words to students with learning disabilities. Learn. Disabil. Res. Pract. **9**(3), 136–141 (1994)
16. Hollier, S., Murray, I.: The evolution of e-inclusion. In: Impagliazzo, J., Lee, J.A.N. (eds.) History of Computing in Education. IAICT, vol. 145, pp. 123–131. Springer, New York (2004). https://doi.org/10.1007/1-4020-8136-7_13
17. Hulme, C., Snowling, M.J.: Reading disorders and dyslexia. Curr. Opin. Pediatr. **28**(6), 731 (2016)
18. Ito, K., et al.: The LJ speech dataset (2017)
19. Karovska Ristovska, A., Maja, F.: Fonts for improvement of the reading abilities in persons with dyslexia. Ann. Fac. Philos. **71**, 437–455 (2018)
20. Kim, W., Linan-Thompson, S., Misquitta, R.: Critical factors in reading comprehension instruction for students with learning disabilities: a research synthesis. Learn. Disabil. Res. Pract. **27**(2), 66–78 (2012)
21. Kirkpatrick, A., Connor, J.O., Campbell, A., Cooper, M.: Web content accessibility guidelines (WCAG) 2.1 (2018). Accessed 31 July 2018
22. Knight, C.: What is dyslexia? An exploration of the relationship between teachers' understandings of dyslexia and their training experiences. Dyslexia **24**(3), 207–219 (2018)
23. Lillo-Martin, D.C., Gajewski, J.: One grammar or two? Sign languages and the nature of human language. WIREs Cogn. Sci. **5**(4), 387–401 (2014). https://doi.org/10.1002/wcs.1297. https://onlinelibrary.wiley.com/doi/abs/10.1002/wcs.1297
24. Marinus, E., Mostard, M., Segers, E., Schubert, T.M., Madelaine, A., Wheldall, K.: A special font for people with dyslexia: does it work and if so, why? Dyslexia **22**(3), 233–244 (2016)

25. Masulli, F., Galluccio, M., Gerard, C.L., Peyre, H., Rovetta, S., Bucci, M.P.: Effect of different font sizes and of spaces between words on eye movement performance: an eye tracker study in dyslexic and non-dyslexic children. Vision. Res. **153**, 24–29 (2018)
26. Mishev, K., Karovska Ristovska, A., Trajanov, D., Eftimov, T., Simjanoska, M.: Makedonka: applied deep learning model for text-to-speech synthesis in macedonian language. Appl. Sci. **10**(19), 6882 (2020)
27. Montali, J., Lewandowski, L.: Bimodal reading: benefits of a talking computer for average and less skilled readers. J. Learn. Disabil. **29**(3), 271–279 (1996)
28. Oberembt, M.: The effects of text-to-speech on students with reading disabilities (2019)
29. Parr, J.M., Fung, I.Y.: A review of the literature on computer-assisted learning, particularly integrated learning systems, and outcomes with respect to literacy and numeracy. Auckland Uniservices, University of Auckland (2000)
30. Paulesu, E., et al.: Dyslexia: cultural diversity and biological unity. Science **291**(5511), 2165–2167 (2001)
31. Pires, G., Nunes, U., Castelo-Branco, M.: Evaluation of brain-computer interfaces in accessing computer and other devices by people with severe motor impairments. Procedia Comput. Sci. **14**, 283–292 (2012). https://doi.org/10.1016/j.procs.2012. 10.032
32. Ping, W., et al.: Deep voice 3: scaling text-to-speech with convolutional sequence learning. arXiv preprint arXiv:1710.07654 (2017)
33. Ribeiro, F., Florêncio, D., Zhang, C., Seltzer, M.: Crowdmos: an approach for crowdsourcing mean opinion score studies, pp. 2416–2419 (2011)
34. Silió, M.C., Barbetta, P.M.: The effects of word prediction and text-to-speech technologies on the narrative writing skills of hispanic students with specific learning disabilities. J. Spec. Educ. Technol. **25**(4), 17–32 (2010)
35. Smith, D.I., Sevensma, K., Terpstra, M., McMullen, S.: Digital Life Together: The Challenge of Technology for Christian Schools. Wm. B. Eerdmans (2020)
36. Snoddon, K.: Wendy Sandler & Diane Lillo-martin, Sign Language and Linguistic Universals. Cambridge University Press, Cambridge (2006). pp. xxi, 547. pb $45.00. Lang. Soc. **37**(4), 628 (2008). https://doi.org/10.1017/S0047404508080883
37. Stodden, R.A., Roberts, K.D., Takahashi, K., Park, H.J., Stodden, N.J.: Use of text-to-speech software to improve reading skills of high school struggling readers. Procedia Comput. Sci. **14**, 359–362 (2012)
38. Stoll, S., Camgoz, N.C., Hadfield, S., Bowden, R.: Text2sign: towards sign language production using neural machine translation and generative adversarial networks. Int. J. Comput. Vis. **128**(4), 891–908 (2020)
39. Viswanathan, M., Viswanathan, M.: Measuring speech quality for text-to-speech systems: development and assessment of a modified mean opinion score (MOS) scale. Comput. Speech Lang. **19**(1), 55–83 (2005)
40. Wery, J., Diliberto, J.: The effect of a specialized dyslexia font, OpenDyslexic, on reading rate and accuracy. Ann. Dyslexia **67**, 114–127 (2017)
41. Wood, S.G., Moxley, J.H., Tighe, E.L., Wagner, R.K.: Does use of text-to-speech and related read-aloud tools improve reading comprehension for students with reading disabilities? A meta-analysis. J. Learn. Disabil. **51**(1), 73–84 (2018)

Serverless Platforms Performance Evaluation at the Network Edge

Vojdan Kjorveziroski$^{(\boxtimes)}$ ⬤, Sonja Filiposka ⬤, and Vladimir Trajkovik ⬤

Faculty of Computer Science and Engineering, Ss. Cyril and Methodius University,
Rugjer Boshkovikj 16, 1000 Skopje, North Macedonia
{vojdan.kjorveziroski,sonja.filiposka,trvlado}@finki.ukim.mk

Abstract. Emerging computer paradigms aim to fulfill the ever-present ideal of running as many applications on existing infrastructure, as efficiently as possible. One such novel concept is serverless computing which abstracts away infrastructure management, scaling and deployment from developers, allowing them to host function instances with granular responsibilities. However, faced with the meteoric growth in the number of IoT devices, the cloud is no longer suitable to meet this demand and a shift to edge infrastructures is needed, providing reduced latencies. While there are existing serverless platforms, both commercial and open-source that can be deployed at the edge, a comprehensive performance analysis is needed to determine their advantages and drawbacks, define open-issues, and identify areas for improvement. This paper analyses three different serverless edge platforms with the help of an existing serverless test suite, outlining their architecture, as well as execution performance in both sequential and parallel invocation scenarios. Special focus is paid to solutions that can be deployed in a standalone fashion, without complex clustering requirements. Results show that while the serial execution performance is comparable among the analyzed platforms, there are noticeable differences in cases of concurrent executions.

Keywords: Serverless computing · Function-as-a-Service · Edge computing · Performance comparison

1 Introduction

Recent breakthroughs in both computer hardware and networking have led to a dramatic increase in the number of new devices [1], allowing novel use-cases, not possible before. Many of these devices interact with the nearby environment and other equipment within it, either because of an operator's command or upon an occurrence of a given event. To ensure a good user experience, such event-driven communication needs low latency [2], a requirement which is challenging to fulfill in the traditional cloud-based architectures. As a result, recently, there has been a gradual shift to the edge of the network [3], closer to the end devices and their users, thus ensuring seamless real-time communication. However, in such scenarios, the question of infrastructure management, execution performance, and application deployment arises, as a result of the more decentralized architecture.

© Springer Nature Switzerland AG 2022
L. Antovski and G. Armenski (Eds.): ICT Innovations 2021, CCIS 1521, pp. 160–172, 2022.
https://doi.org/10.1007/978-3-031-04206-5_12

A possible solution that would ease the use of such edge-based infrastructure is serverless computing [4], with its function-as-a-service semantics [5], allowing developers to write granular functional elements, which are easier to create, maintain, and scale. Cloud-based serverless solutions have existed for a number of years [6] and have proven very popular among businesses and developers alike, reducing the time to market, and providing a more cost-effective alternative to the more traditional Platform-as-a-Service (PaaS) or even Infrastructure-as-a-Service (IaaS) options of application deployment.

Taking into account that serverless at the edge is still a novel topic, there are already some open-source and commercial solutions, with a common aim of simplifying the infrastructure management and function deployment processes. However, their implementation, developer interfaces, and supported technologies vary significantly, and this lack of interoperability and standardized access methods is one of the main issues of serverless computing today. One consequence of this diversity is the difference in performance offered by the various serverless platform implementations, an aspect of paramount importance in environments requiring low response times, such as at the edge of the network.

The aim of this paper is to determine the performance characteristics of popular serverless platforms at the edge, both commercial and open source, by utilizing an existing set of benchmarks dedicated to serverless computing and adapting them to the platforms at hand.

The rest of the paper is structured as follows: in Sect. 2 we outline related work to this topic and recent notable efforts of characterizing serverless execution performance in different environments, among different platforms. In Sect. 3, we describe the employed methodology, the platform selection process, as well as the set of benchmarks being used. We then proceed to present the acquired results in Sect. 4, analyzing the performance differences between the platforms. We conclude the paper with Sect. 5, outlining plans for future work.

2 Related Work

With the rise in popularity of serverless computing, there has been an increasing interest from both the academic community, as well as the industry for the development of new solutions. These novel platforms tackle different aspects of the associated open issues with serverless computing [7], ranging from execution efficiency, to ease of use. Many academic implementations also show benchmark results which quantitively compare their improved performance to other popular serverless platforms, either at the edge [8], the cloud [9] or in a hybrid hierarchical model [10], depending on the targeted execution location of the platform itself. However, the lack of standardized tests which would encompass all different aspects of the implementation at hand, and would facilitate easier results comparison, has led to the development of dedicated benchmark suites.

The authors of [11] present FunctionBench, a set of benchmark tools for cloud-based serverless platforms whose aim is to characterize network, disk, and compute performance of the targeted platforms. Similar to this, Das et al. [12] focus on the edge counterparts of these public cloud platforms, which can be deployed on private infrastructure, thus remotely orchestrating the deployment of new functions. Their results show that AWS Greengrass is more efficient in such constrained environments compared to Azure IoT hub, another commercial product for serverless computing at the edge.

Gorlatova et al. in [13] aim to bridge this gap, comparing both edge-based and cloud-based solutions, confirming that the cold start problem [14], experienced during the initial startup of a function after it has been scaled down to zero [15] plays a major role in functions' execution time. Finally, the authors of [16], unlike previous efforts, focus solely on open-source solutions, and their performance characteristics.

While these works offer an important insight into the different available serverless implementations, they focus solely on either commercial products or open-source ones, without thorough comparison between them. To the best of our knowledge no performance analysis focusing on both spectrums is currently available.

3   Methodology

Measuring the performance of different execution environments requires standardized tests based on common utilities that can be executed across different platforms. For this reason, we have decided to reuse the test suite presented in [11], whose code has been open-sourced and made publicly available [17]. Nevertheless, the lack of consistent programming interfaces between the different platforms required manual changes to the benchmarks, which have been initially developed for cloud-based serverless platforms.

In the subsections that follow we explain in detail the platform selection criteria, describe the different tests that have been utilized, and present the infrastructure where they have been executed.

3.1   Platform Selection

The primary criteria for choosing which platforms to include was the requirement for standalone deployment, without the need for multiple worker machines, or clustering setups. This led to the omission of many popular open-source solutions, such as Openwhisk[1] and OpenFaaS[2] since they rely on complex container orchestration platforms such as Kubernetes or Docker Swarm. Nonetheless, in our opinion, keeping the infrastructure as simple as possible, and deployable in severely constrained environments is an important aspect of edge architecture design, justifying the strict selection criteria.

We have included FaasD [18, 19], which is a lightweight version of OpenFaaS, whose primary purpose is deployment on edge devices with limited resources, thus completely fulfilling the above criteria. As a relatively new open-source solution, it is being actively developed, and has not been included in any of the previously mentioned platform evaluations, a fact that has further contributed to its selection.

FaasD is not the only scaled-down version of a popular serverless platform. There have been efforts to simplify the requirements of OpenWhisk in the past as well, resulting in OpenWhisk-Light [20], but unfortunately it is no longer actively maintained, making it an unsuitable choice for inclusion.

---

[1] https://openwhisk.apache.org/ [Online] (Accessed: 31.05.2021).

[2] https://www.openfaas.com/ [Online] (Accessed: 31.05.2021).

As representatives of the commercial serverless edge solutions, we have selected both AWS Greengrass and Azure IoT Hub, based on their popularity, and the fact that they have been included in other performance studies as well, allowing us to compare the obtained results more easily. Furthermore, the selected benchmark suite has been purposefully developed for the cloud counterparts of these products, providing an opportunity to reflect on the needed changes to adapt the test functions to the edge.

3.2 Test Types

Table 1. Test parameters and description

Test category	Test name	Parameters	Purpose
CPU & Memory	Float Operation	Random number	Common operations
	Matmul	Random number (matrix size)	Matrix multiplication
	Chameleon	500 × 500 (number of columns and rows)	HTML table rendering
	Image Processing	Object storage location	Image transformations
	Linpack	Random number (matrix size)	Linear equation solver
	PyAES	1000, 100 (length & iterations)	AES operations
	Model Training	Object storage location	ML model training
	Video Processing	Object storage location	Video encoding
Network	iperf3	IP address & duration	Network throughput
	JSONDumpsLoads	URL of JSON file[a]	JSON manipulations
	Object Download	Object storage location	Object storage performance
Disk	Dd	100 M, 1 (block size & count)	Dd disk speed
	Random I/O	100, 1024 (file size & byte size)	Python random I/O
	Sequential I/O	100, 1024 (file size & byte size)	Python sequential I/O
	Gzip	50 MB (compression size)	Gzip compression

[a] https://data.parliament.scot/api/departments [Online] (Accessed 31.05.2021).

The testing suite that has been adapted to run on the previously selected serverless edge platforms consists of 15 different functions, divided into three distinct categories: CPU, network, and disk benchmarks. Each of these tests requires different input parameters which represent the input data that is being worked upon. Considering the fact that these benchmarks have been created to test cloud-based platforms, we have replaced all cloud services such as object storage with either native features of the platform itself, or with other locally hosted alternatives. This allows more relevant latency information, where cloud communication can be completely avoided.

The lack of standardized serverless function formats meant that all 15 benchmarks had to be adapted manually to the three different platforms, resulting in 45 total functions. While there are efforts for provider cross-compatibility [10, 21], they are third party solutions that are not officially supported, often leading to a reduced feature set.

Table 1 shows more details about each of the tests, their purpose, as well as any inputs. In terms of the input parameters, to ensure equal testing conditions, all function instances that require an input number have been invoked with the same random number set. The other parameters have been chosen to strike a balance between load stressing and execution time, simulating common operations that might be performed at the network edge.

3.3   Execution Environment and Method

All three platforms have been deployed on $\times 86$ based virtual machines, each allocated with 2 vCPU cores and 4GB of RAM, simulating resource constrained execution devices such as widely available computing boards. Each test has been executed in multiple fashions: i) serially, simulating one request at a time; ii) parallelly, simulating up to 100 requests at a time; iii) serially, but with the function scaled down to 0 replicas to test the cold-start latency. All tests, across all execution scenarios have been executed: 1, 5, 10, 25, 50, 75, and 100 times, thus measuring the total response time, instead of simply the execution latency, which does not take into account network latency. This approach allows us to determine the efficiency of the underlying communication protocol as well, since it differs among the included platforms.

In the following section we present the obtained results, as well as any platform specific information that might have influenced the outcome.

4   Results

The three different platforms vary significantly in their developer interfaces, function provisioning, scaling mechanisms, communication protocols, and overall feature set. Each of them uses different tooling mechanisms to ease the development of new functions and provides a ready-made set of templates. However, one thing that they all have in common is reliance on containerization as a runtime environment for the functions.

4.1   FaasD

FaasD is a new serverless platform created primarily for deployments at the network edge, on resource constrained devices. It is a more lightweight version than OpenFaaS,

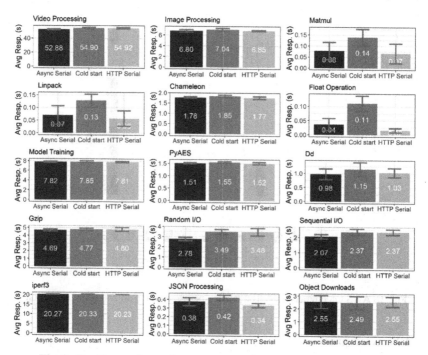

Fig. 1. FaasD execution performance comparison between different modes

thus supporting the same function format. However, to achieve the desired simplicity, some notable features have been omitted, such as automatic scaling of function instances to more than a single replica, or automatic scaling to zero. Unfortunately, because of no container orchestration middleware, multiple function instances are not possible, while manual pausing of idling containers is possible using the application programming interface (API).

All functions are instantiated from manually built Docker images, and generic starter templates for different programming languages are provided. It supports two modes of invocation, either synchronous where the user waits until the response is returned or asynchronous where the serverless platform can issue a callback whenever a function has completed.

Figure 1 shows the average response time incurred during 100 serial invocations with one request at a time of the different tests, using the three invocation methods. The cold start latency is measured by synchronously invoking the function through the HTTP interface. It is evident that in most of the tests there is a cold start delay, a consequence of the paused container which must be resumed.

Parallel execution using asynchronous invocation is not possible in FaasD, since only one request can be processed at a time in this fashion, by a single container. However, this would not be an issue in environments with a container orchestrator, since spawning multiple container instances would solve the problem.

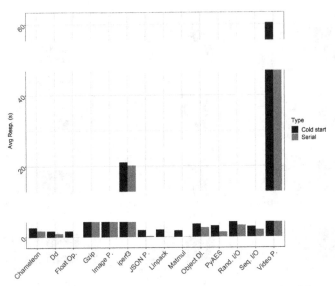

Fig. 2. AWS Greengrass execution latency between cold start and serial execution

4.2 AWS Greengrass

AWS Greengrass is an Amazon Web Services product which allows lambda function deployment on customer owned edge devices. One of the primary features is the cross compatibility with the cloud-based Lambda service, where the same function code can be reused. It supports two modes of execution, either native or containerized, with options of strictly limiting the execution time, and memory consumption of the function itself, similar to the cloud counterpart of the service. Support for scaling to zero of unused functions as well as dynamically increasing the number of replicas in response to user demand is also supported, and transparent to the operator of the device. As a result of this, no request level isolation is guaranteed and multiple subsequent requests could be executed in the same container, sharing leftover temporary files.

The primary communication mechanisms with other devices and between Lambda functions is by message passing, where either the on-premise Greengrass device can be used as a message broker, or the cloud.

Unlike other serverless platforms, Greengrass does not rely on Docker images as a distribution format, instead all Lambdas are packaged as archive files, with a strict size limit imposed. This prevented the execution of the Model Training benchmark, whose resulting package was simply too large for this platform, explaining its omission from subsequent figures related to Greengrass.

Figure 2 shows a comparison between the regular latency and the cold-start latency during 100 invocations of all functions in the test suite. Similarly to FaasD, a cold-start delay is present, something that has been confirmed by other benchmarks focusing on commercial edge platforms as well [13].

4.3 Azure IoT Hub

Azure IoT Hub is a Microsoft service which supports function deployments to customer-owned edge devices as well. Like the other platforms, it relies on containerization for function execution, using the native Docker image format. There are no requirements to what interfaces the function must implement to be run in the serverless environment, and it is left to the developer to devise the communication mechanisms. Of course, message passing is one of the supported communication options, and in a similar fashion to Greengrass, the messages can either be routed locally or via the cloud. Unfortunately, neither scaling down to zero is supported nor horizontal scaling to multiple function instances. As a result of this, any concurrency is left to be implemented by the developers and built into the functions themselves, a popular option being the introduction of threading.

Cross-compatibility with the cloud counterpart of the serverless service is only possible when the functions are developed in the C# programming language, in all other instances, they must be packaged as Docker container by the users themselves. However, there is an option to run local instances of the Azure blob storage service and the relational database service, with the same API as those hosted in the cloud, thus removing the need for hosting on-premise alternatives.

There are no cold-start delays during the function invocation, since scale-down-to-zero is not supported, and the container instance hosting the function is always left in a running state.

4.4 Performance Comparison

Even though all selected platforms fulfill the primary goal of hosting serverless functions at the network edge, they vastly differ in their implementations. In this subsection we compare both their performance, and parallelization limits.

Synchronous Serial Execution

Figure 3 shows the serial synchronous execution performance of all platforms, across all different tests. In terms of FaasD, the HTTP serial synchronous execution is shown. As mentioned previously, results for the Model Training benchmark are absent for the Greengrass platform, since the resulting function package was too large to be deployed. This is a known problem in serverless environments, and one possible option is library size optimization through the removal of unused code [22, 23]. The results show that Greengrass exhibits the best performance for I/O bound workloads, such as the Dd, sequential I/O, random I/O and Gzip compression tests. A noticeable performance difference is seen in the model training and the float operation benchmarks, with the best performing platform being 25.76 and 80 per cent faster, respectively, than IoT Hub. All other tests show comparable levels of performance, with Greengrass being the fastest option in 12 test instances, and FaasD in 3, out of 15.

Parallel Execution

Comparing the parallel execution response time among the different platforms is challenging, because of the different levels of support for function parallelism. Both FaasD

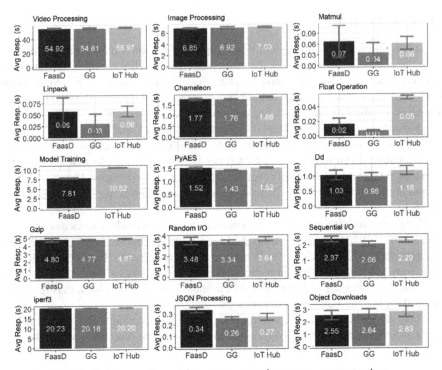

Fig. 3. Serial execution performance comparison, one request at a time

and Azure IoT Hub do not natively support automatic function scaling, a feature that is present in Greengrass. Furthermore, since all FaasD functions were implemented using the official Python template which includes a web server, the number of parallel requests that can be processed depends on the number of worker threads that have been spawned. If this value is left to its default settings, it is calculated based on the number of CPU cores that are available on the machine. Finally, in the case with IoT Hub, the developers have full control over any parallelization aspects.

Figure 4 shows the parallel execution performance of 14 out of the 15 total tests across all environments, where the maximum response time in seconds is given for 1, 5, 10, 25, 50, 75, and 100 parallel invocations. The iPerf3 test is omitted since it is not relevant in this case – iPerf3 supports only a single concurrent connection on a given port and conducting multiple tests at the same time would have required spawning of numerous parallel iPerf3 instances on the target machine. It is evident that a noticeable difference exists between the execution latencies of the platforms, and as expected, this is as a result of the varying levels of concurrency supported by each of them, as discussed in the previous subsections. The I/O advantage of Greengrass is also noticeable during parallel execution as well, while the limited parallelism of FaasD and IoT Hub 4 allow them to finish all tasks in a reliable manner, albeit, in some cases, slower. At the end, the concurrency level of each task should be tweaked as per its resource requirements, to mitigate situations which would lead to premature resource exhaustion, as is the case with the image processing benchmark for both Greengrass and IoT Hub with 100

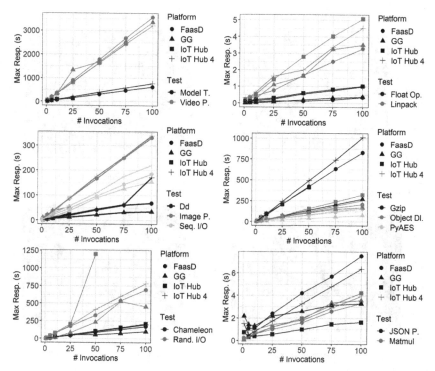

Fig. 4. Parallel execution performance comparison with varying degree of concurrency

concurrent workers, where 50 and 25 parallel executions, respectively, were not feasible. Total response time should be considered as well during such tweaking, which would prevent drastic latency increase during heavy I/O tasks, as is the case with IoT Hub and the Random I/O test.

With the aim of further investigating the maximum number of concurrent tasks executed by each platform, we have plotted the response times using histograms, as shown in Fig. 5. To normalize the different results among the different platforms, each response time during a parallel execution has been divided by the average response time during the serial, sequential, execution of the task, and the bin size has been set to the ratio between the average execution time during a parallel execution and the average execution time during a serial, sequential, execution. This ratio can also be seen as a slowdown coefficient between the parallel and sequential executions due to the higher concurrency. It is worth noting that the average response time takes into account any incurred communication latency, while the execution time relates only to the time taken to complete the task once it has been invoked, ignoring any communication delays.

By utilizing the previously described approach, we have confirmed that the FaasD function execution environments within the containers have defaulted to 4 worker instances when invoked parallelly using the synchronous HTTP interface, and that FaasD async execution supports only a single concurrent request per container instance. As a result of this, and exploiting the greater control that IoT Hub offers in terms of parallelism,

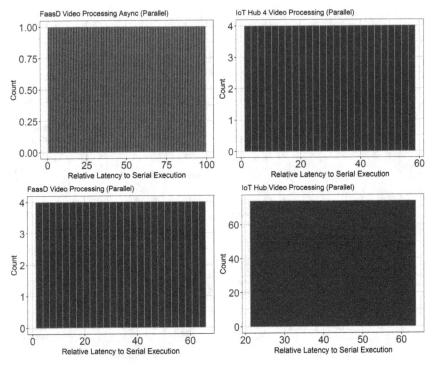

Fig. 5. Number of concurrent worker processes during parallel execution

we have decided to execute all tests on this platform using two levels of concurrency – 100 and 4. In this way, results that relate to the higher concurrency rating can be compared to Greengrass, and those with concurrency of 4 to FaasD. Unfortunately, some tests led to premature resource exhaustion on the test machines and could not be completed with all concurrency targets, explaining their omission in Fig. 4, further validating the point in terms of careful resource allocation, discussed earlier. These results have been used to show that, indeed, the manual concurrent implementation of the functions in Azure IoT Hub, using a maximum of 100 parallel workers, has met the desired concurrency level, as per the histogram shown in the bottom right corner in Fig. 5, analyzing the concurrency rating of the video processing test with 75 parallel requests issued.

5   Conclusion and Future Work

Using a set of existing benchmarks for cloud-based serverless solutions, and through their adaptation for edge-based platforms, we have compared the execution performance and response times of three different serverless products that can be deployed at the network edge. By including both open-source and commercial software, we have bridged the gap of other similar contributions, where the focus has been given exclusively to either option. In total, 15 different tests have been adapted, divided into three distinct categories, and each of these tests has been invoked in multiple scenarios, simulating both parallel and sequential task execution.

The results show that there is a comparable difference between the platforms when it comes to sequential, one at a time execution of functions, but a widening gap in terms of parallel execution. The main reasons for this are the varying levels of concurrency that each platform offers, with some choosing to abstract all scaling away from the developers, such as AWS Greengrass, while others catering only to the function deployment, and leaving concurrency to the implementation of the function itself, for example IoT Hub and to some extent FaasD. This varying rate of concurrency also impacts the execution performance and reliability, since all scaling must be done in respect to the available system resources, eliminating situations where tasks fail due to lack of disk space or memory.

When it comes to overall performance, AWS Greengrass shows a noticeable advantage in I/O related benchmarks, both in terms of serial and parallel execution, while FaasD offers more diverse execution options than the alternatives, supporting different communication protocols, and invocation strategies. The advantages of FaasD also extend to its observability, and integration options with other popular monitoring tools, instead of relying on similar options from commercial vendors. Finally, IoT Hub offers a more hands-on approach where the developer is responsible for the concurrency of the functions being executed.

Through the analysis of the test results, we have further confirmed some of the existing open issues in serverless architectures, such as the cold-start delay, non-standardized APIs [7], and function size ballooning due to many dependencies [14, 15]. It is worth noting that serverless computing is still an area under active research and it is expected to further grow in the coming years, as solutions to the open problems are found [24].

In the future, we plan to extend our research of serverless platforms' performance with the inclusion of additional solutions that use alternative execution environments and do not focus on containers, thus helping to eliminate some of the issues related to containerization, such as the cold-start latency, overlay file system performance, and platform cross-compatibility.

References

1. Number of connected devices worldwide 2030. Statista. https://www.statista.com/statistics/802690/worldwide-connected-devices-by-access-technology/. Accessed 01 June 2021
2. Pfandzelter, T., Bermbach, D.: IoT data processing in the fog: functions, streams, or batch processing? In: 2019 IEEE International Conference on Fog Computing (ICFC), pp. 201–206. Prague, Czech Republic (2019). https://doi.org/10.1109/ICFC.2019.00033
3. Bittencourt, L., et al.: The internet of things, fog and cloud continuum: integration and challenges. Internet Things 3–4, 134–155 (2018). https://doi.org/10.1016/j.iot.2018.09.005
4. Kratzke, N.: A brief history of cloud application architectures. Appl. Sci. 8(8), 1368 (2018). https://doi.org/10.3390/app8081368
5. El Ioini, N., Hästbacka, D., Pahl, C., Taibi, D.: Platforms for serverless at the edge: a review. In: Zirpins, C., et al (eds.) Advances in Service-Oriented and Cloud Computing, vol. 1360, pp. 29–40. Springer, Cham (2021). https://doi.org/10.1007/978-3-030-71906-7_3. Accessed 09 April 2021
6. Introducing AWS Lambda. Amazon Web Services, Inc. https://aws.amazon.com/about-aws/whats-new/2014/11/13/introducing-aws-lambda/. Accessed 01 June 2021

7. Hellerstein, J.M., et al.: Serverless Computing: One Step Forward, Two Steps Back (2018). http://arxiv.org/abs/1812.03651. Accessed 05 Feb 2021

8. Wolski, R., Krintz, C., Bakir, F., George, G., Lin, W.-T.: CSPOT: portable, multi-scale functions-as-a-service for IoT. In: Proceedings of the 4th ACM/IEEE Symposium on Edge Computing, pp. 236–249. Arlington Virginia (2019). https://doi.org/10.1145/3318216.336 3314

9. Ling, W., Ma, L., Tian, C., Hu, Z.: Pigeon: a dynamic and efficient serverless and FaaS framework for private cloud. In: 2019 International Conference on Computational Science and Computational Intelligence (CSCI), pp. 1416–1421. Las Vegas, NV, USA (2019). https://doi.org/10.1109/CSCI49370.2019.00265

10. Huang, Z., Mi, Z., Hua, Z.: HCloud: a trusted JointCloud serverless platform for IoT systems with blockchain. China Commun. 17(9), 1 (2020). https://doi.org/10.23919/JCC.2020.09.001

11. Kim, J., Lee, K.: FunctionBench: a suite of workloads for serverless cloud function service. In: 2019 IEEE 12th International Conference on Cloud Computing (CLOUD), pp. 502–504. Milan, Italy (2019). https://doi.org/10.1109/CLOUD.2019.00091

12. Das, A., Patterson, S., Wittie, M.: EdgeBench: benchmarking edge computing platforms. In: 2018 IEEE/ACM International Conference on Utility and Cloud Computing Companion (UCC Companion), pp. 175–180. Zurich (2018). https://doi.org/10.1109/UCC-Companion. 2018.00053

13. Gorlatova, M., Inaltekin, H., Chiang, M.: Characterizing task completion latencies in multi-point multi-quality fog computing systems. Comput. Netw. 181, 107526 (2020). https://doi.org/10.1016/j.comnet.2020.107526

14. Gill, S.S., et al.: Transformative effects of IoT, blockchain and artificial intelligence on cloud computing: evolution, vision, trends and open challenges. Internet Things 8, 100118 (2019). https://doi.org/10.1016/j.iot.2019.100118

15. Gadepalli, P.K., Peach, G., Cherkasova, L., Aitken, R., Parmer, G.: Challenges and opportunities for efficient serverless computing at the edge. In: 2019 38th Symposium on Reliable Distributed Systems (SRDS), pp. 261–2615 (2019). https://doi.org/10.1109/SRDS47363.2019. 00036

16. Palade, A., Kazmi, A., Clarke, S.: An evaluation of open source serverless computing frameworks support at the edge. In: 2019 IEEE World Congress on Services (SERVICES), vol. 2642–939X, pp. 206–211 (2019). https://doi.org/10.1109/SERVICES.2019.00057

17. kmu-bigdata/serverless-faas-workbench. BigData Lab. in KMU (2021). https://github.com/kmu-bigdata/serverless-faas-workbench. Accessed 27 April 2021

18. openfaas/faasd. OpenFaaS (2021). https://github.com/openfaas/faasd. Accessed 09 May 2021

19. faasd - OpenFaaS. https://docs.openfaas.com/deployment/faasd/. Accessed 31 May 2021

20. Kravchenko, P.: kpavel/openwhisk-light (2020). https://github.com/kpavel/openwhisk-light. Accessed 09 May 2021

21. The Serverless Application Framework|Serverless.com. serverless. https://serverless.com/. Accessed 27 April 2021

22. Christidis, A., Davies, R., Moschoyiannis, S.: Serving machine learning workloads in resource constrained environments: a serverless deployment example. In: 2019 IEEE 12th Conference on Service-Oriented Computing and Applications (SOCA), pp. 55–63. Kaohsiung, Taiwan (2019). https://doi.org/10.1109/SOCA.2019.00016

23. Pelle, I., Czentye, J., Doka, J., Kern, A., Gero, B.P., Sonkoly, B.: Operating latency sensitive applications on public serverless edge cloud platforms. IEEE Internet Things J. 8(10), 7954–7972 (2021). https://doi.org/10.1109/JIOT.2020.3042428

24. Varghese, B., Buyya, R.: Next generation cloud computing: new trends and research directions. Future Gener. Comput. Syst. 79, 849–861 (2018). https://doi.org/10.1016/j.future.2017. 09.020

Sensor Systems, IoT

Integration of IoMT Sensors' Data from Mobile Applications into Cloud Based Personal Health Record

Snezana Savoska[1]([✉]) [iD], Natasha Blazeska-Tabakovska[1] [iD], Ilija Jolevski[1] [iD],
Andrijana Bocevska[1] [iD], Blagoj Ristevski[1] [iD], Vassilis Kilintzis[2] [iD], Vagelis Chatzis[2],
Nikolaos Beredimas[2], Nicos Maglaveras[2] [iD], and Vladimir Trajkovik[3] [iD]

[1] Faculty of Information and Communication Technologies – Bitola, University "St. Kliment Ohridski" – Bitola, ul. Partizanska bb, 7000 Bitola, Republic of North Macedonia
{snezana.savoska,natasa.tabakovska,ilija.jolevski,
andrijana.bocevska,blagoj.ristevski}@uklo.edu.mk
[2] Laboratory of Computing, Medical Informatics and Biomedical Imaging Technologies, Aristotle University, Thessaloniki, Greece
{billyk,nicmag}@med.auth.gr, beredim@auth.gr
[3] Faculty of Computer Science and Engineering, "Ss. Cyril and Methodius" University Skopje, Skopje, Macedonia
vladimir.trajkovik@finki.ukim.mk

Abstract. Mobile applications for vital signs measurement are popular in ambient assisted living environments. They typically include various wearables and devices with their applications connected with the production company with some security preferences. When a manufacturer develops and markets a specific sensor, a smartphone application is developed. Many of the sensors and devices are secured and can work only with its host application developed by the manufacturer. However, for these sensors and devices related to a patient's cloud-based Personal Health Record (PHR), different mobile applications must be created to connect all additional wearables or IoT-based medical sensors or devices to the appropriate cloud-based PHR. All data from these mobile applications needs be collected into the PHR database securely and in time to store medical information for the patient's health in the PHR. This paper presents the integration of two types of mobile applications – for medical professionals and citizens with the PHR, intended for the Cross4all project. The paper explains how the mobile application is integrated with different sensors and devices and the cloud PHR. We also describe the challenges that have arisen from the pilot project implementation and solutions. The presented concept improves cross-border evidence-based healthcare and integrates the e-prescription and e-referral system intended to solve some cross-border healthcare problems for foreign citizens.

Keywords: e-Health · Health information systems · Mobile applications · IoMT · PHR

L. Antovski and G. Armenski (Eds.): ICT Innovations 2021, CCIS 1521, pp. 175–187, 2022.
https://doi.org/10.1007/978-3-031-04206-5_13

1   Introduction

The current large amount of IoMT-based devices, including wearables and sensors for remote patient monitoring, are designed in order to increase efficiencies, decrease care costs, and improve outcomes in healthcare. They also tend to improve e-health services by improving the possibilities for medical care in the distance, especially in a pandemic situation. Healthcare organizations are increasingly leveraging the potential of IoT and IoMT (Internet of Medical things) with their ability to collect, store, analyze, transmit and use health data. Thus, the IoMT has a vital role in personal-EHRs.

When we are talking about the PHRs, we can say that the health-related data can be collected from various sources. The sources can be EHRs, patient's records from health providers as prescription data, labs' data, bio-monitoring data, referral data, or directly from a sensor for vital signs measurements or some general-purpose data as exercise habits, diet statistics, and food or some screening or exposome data, etc. [1]. The cloud-based personal health record (PHR) concept supports collecting data as scanned medical documents for the patients, information beyond the patient's medical history, such as data related to conducting healthy life, style of living, and living conditions. These data can form a more holistic representation of the person's health lifecycle, and covering all kinds of physical, psychological, and social aspects. In this cloud-based PHR-based approach, the records are fully owned by the patient [2], controlled and secured simultaneously [3].

The Cross4all project concept combines the power of cloud computing as well as service-oriented architectures. It provides patient-centric care, including telecare possibilities, self-care elements, and increasing evidence-based healthcare management in a cross-border environment. The project includes several mobile applications for medical professionals and patients, focusing on serving the needs of the older people, people with disabilities [2] and geographically or socially isolated persons. This integrated Cross4all ecosystem with applications and many features aims to provide a wide range of capabilities. When the patient decides to create their PHR on the project, the patient becomes a PHR owner. The patient then can use the web platform through his/her device and store, view and use their health profile and history, medication plan, visits to health service providers, and all according to FHIR standards [2, 3]. A patient can also upload additional medical-related documents, such as data obtained from a mobile application for vital sign measurements and connected sensors or medical devices. The patient can also find, select, and share data from his/her PHR with selected cross-border healthcare professionals. The selected medical professionals can monitor patients' health, give advice, prescribe some drugs through an e-prescription system, or create e-referral for specialists of other medical institutions. The patient can grant temporary permission to selected medical persons or pharmacists to gain medical care.

The paper explains the real pilot of project Cross4all. The pilot implements cross-border cloud-based PHR and uses mobile applications for medical professionals and citizens in the real environment in Greece and R. of North Macedonia. Many challenges and possible solutions were taken into consideration during the process of implementation of the Cross4all pilot. The second section provides an overview of recent related works, while the third section highlights some points of the deployment of medical devices and sensors through the usage of mobile applications for patients and professionals and their connection with the cloud PHR. The fourth section emphasizes some aspects of data acquisition from devices and storing them into cloud PHR. The actual data acquisition is discussed in the fifth section. In the conclusion section, directions for improving the quality of implementation and proposals for future work are given.

2 Related Work

Many researchers consider IoT and IoMT challenges in the last decade, focusing their research on different issues. Some of them consider end-user privacy and security challenges for IoT healthcare applications [17]. They consider IoT its layered architecture, describe privacy and security services, providing to view confidentiality as privacy and security intersection. Some authors [4] analyze the ideas as well as the impacts of IoT on the new e-health solutions designing and identifying the main challenges that determine successful IoMT-based e-health system adoption.

A general overview of IoT in healthcare medical systems is presented in [5]. It discusses applications connected with IoMT as well as medical data analysis. In addition, the advantages of IoMT in the healthcare sector is discussed. IoT-related threats, some issues of factors determining the future trends of IoT in medical and healthcare systems are also taken into consideration. The role of IoT as well as the role of other interdisciplinary fields are considered in the paper in order to boost smart and pervasive healthcare. An automated telehealth monitoring system for monitoring and measuring different physiological parameters of the body using Arduino is presented in [6]. IoT devices are proposed to collect and store the needed parameters as well as evaluate the data gained from the IoT devices. Data are sent and stored to the cloud system, in database in PHC. In case of emergency, the tele-ambulance is called and it is integrated with all medical facilities in disposal, to save lives in emergencies. At the same time, there is no time to take the patient to the main hospital. The paper [7] surveys various usages of IoT medical and healthcare technologies, reviewing the state of the art services, applications, recent trends in IoT-based medical and healthcare solutions. Also, the paper considers various challenges posed, including security and privacy issues, service providers, and end-users experiences. In addition, the paper takes into consideration some innovative IoT-enabled technologies as big data, cloud computing, block chains, fog computing and others in order to leverage modern medical and healthcare facilities as well as to mitigate healthcare resources' burden.

The authors of the paper [8] have identified key components of the end-to-end IoT healthcare system and propose a wider healthcare model that could be applied to a wide range of IoT-based healthcare systems. The focus is on monitoring sensors for various healthcare parameters, cloud technologies as well as long and short-range communications standards. The original contribution is made by focusing on Low-Power Wide-Area Networks (LPWANs), highlighting their unique suitability for use in IoT systems. Paper [9] considers how promising technologies such as ambient assisted living, cloud computing, wearables and big data that are being applied in medicine and healthcare industry. In addition, in the paper various regulations for IoT and e-health regulations are taken into consideration as well as some policies worldwide in order to determine how they can assist in the IoT and cloud computing's sustainable development in the healthcare industry.

The analysis and a systematic review of the future technologies instrumental for a large-scale of the Human H-IoT systems development are presented in [10]. This paper identifies also how these new technologies are used in the future course with aim to improve the Quality of Service (QoS). The paper [16] survey IoT security and privacy issued and pointed out of some state-of-the-art solutions that can be taken as a practical guide for how to develop a modern IoT or IoMT application. Some aspects of a circular economy in the healthcare domain are also considered in the paper. Some points of how to create the mechanisms for protection of data that have to be acquired from the devices to the cloud ends and how to process, transmit and store them, in order to be reused, are also given in the paper.

A taxonomy for mobile health applications is described in [12]. The taxonomy attempts to classify applications that have to be controlled by a healthcare facility but it has to be used by the medical practitioners for healthcare-specific data exchange as well as to provide healthcare and medical communications. The important issue is to present to interested data custodians, as consultants and auditors who are responsible for overseeing health information security. This is just a brief holistic review of statements of used technologies for the project.

3   Mobile Applications Connected with Devices or Sensors for Vital Signs Measurement

The project Cross4all uses the potential of the IoT concept together with the PHR concept to increase the number of people that access high-quality health and social services in the cross-border area, thus promoting safe aging, early prevention, and independent living for all [2, 11].

Two categories of health-related mobile applications were developed in the Cross4all project: one for citizens and one for professionals (medical practitioners). According to the findings presented in [12], they support personalized healthcare and remote monitoring and provide support to clinical information management. The first category of this kind of application communicates with multiple personal healthcare devices that are user-friendly [13]. The application supports remote monitoring and measuring of patients' vital signs. This application saves data into PHR with user involvement [14]. The collected data is being used and evaluated by the users or by medical professionals. A mobile device/tablet and internet connection are involved in exchanging collected data between patients and medical providers. The second category of mobile applications is those used by medical professionals. Both applications are based on the Android operating system. The Android operating system is chosen for several major differences (and in our case, advantages) rather than iOS. Firstly, Android enables more low-level access to the Bluetooth communications stack than iOS. Secondly, the distribution and installation of the Android apps are far more straightforward than the distribution of iOS apps.

4 Installation and Integration of Mobile Applications with the Cloud-Based PHR

The two versions of mobile applications have been developed. Mobile application for patients uses day-to-day monitoring devices such as blood pressure monitor, blood glucose monitor, weight scale, pulse oximeter etc. Mobile application for the medical professionals and the previous devices uses more complex medical devices and sensors, such as digital stethoscope, Bluetooth ECG. Both applications need permissions to access the Bluetooth and Location services of the Android OS. They used their native application integrated with the backend - webPHR cloud application and provide saving data into patient's PHR.

The integration of the mobile applications with the underlying backend consists of communication with two distinct services: (1) Authentication sub-system, which is OpenID Connect, based Keycloak server, and (2) Backend API service that is storing the medical health data for the PHR. Since the location of the services is not hardcoded in the applications themselves, to support changes between staging and production environments, the only thing left to configure is the URL locations of the Keycloak services and the URL of the backend API that is done during the first start of the apps.

The registration of new users in the PHR Cloud is supported through the Keycloak Web Registration form opened directly in the smartphone application, so no other device is needed to enroll new users (patients) in the Cloud PHR.

The medical professionals have to use a separate application for each device (Fig. 1a). The mobile applications are connected with the Cross4all mobile application. The patients used one application with many modules (Fig. 1b).

Shared with me › Cross4All ·			
Name ↓	Last modified	File size	
Stethoscope3MService.apk	Mar 19, 2021 Evangelos Chatzis	4 MB	
ShimmerECGService.apk	Jan 14, 2021 Evangelos Chatzis	10 MB	
Observations.apk	Feb 22, 2021 Evangelos Chatzis	3 MB	
MIRSpirometerService.apk	Jan 14, 2021 Evangelos Chatzis	5 MB	
MedisanaOximeterService.apk	Jan 14, 2021 Evangelos Chatzis	1 MB	
MedisanaBPMService.apk	Jan 14, 2021 Evangelos Chatzis	1 MB	
ForaThermometerService.apk	Jan 14, 2021 Evangelos Chatzis	1 MB	
FORAGlucoseService.apk	Feb 18, 2021 Evangelos Chatzis	1 MB	
Cross4All.apk	Feb 22, 2021 Evangelos Chatzis	4 MB	
AEGScaleService.apk	Jan 14, 2021 Evangelos Chatzis	1 MB	

Fig. 1a. Main HCI for Mobile applications for medical professionals.

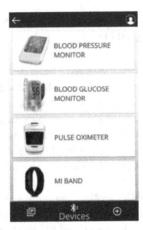

Fig. 1b. Mobile application for the patients.

5   Data Acquisition into the Cloud-Based PHR

The proposed IoT methodology includes cross-border use, with a focus on the needs of older and disabled people [15] as well as geographically and social isolated individuals. This approach provides the data collection, creating a patient database and data analysis for effectual patients' treatment. The proposed workflow processes of the IoMT concept make the connections between healthcare appliances, PHR, Internet, a mobile application for patients and medical professionals, and services using eight steps.

Step 1: Downloading and installing. From available applications (Pulse Oximeter PM150, Blood Pressure Monitor BU550, Glucose Meter Fora Diamond MINI, Ear Thermometer Fora IR20b, Vitalograph COPD-6, Weight Scale AEG PW5653 BT) the medical professionals and patient download and install the mobile application/s on own mobile device.

Step 2: Bluetooth connection. The sensor-based medical device and the mobile device (tablet) communicate through Bluetooth, and connection has to be established. For the ECG and the stethoscope, the user has to pair the device with the tablet. In the tablet's Bluetooth, the setting is input 4-digit pair code.

Step 3: Vital signs measurement. The sensor-based medical device measures the patients' vital signs.

Step 4: Measured data collection. After measuring with the corresponding device and once the measurement is completed with the button "Start a measurement and click here when it is completed" or "press here to link" (for the pulse oximeter) the name of the device will appear on the screen. Communication between devices is established, and the mobile application collects the patients' data from the device.

Step 5: Start communication service. When the device's name appears on the screen (Fig. 2a) with the button "Start Service", the connection can be started with Cross4All Platform. The message "connect with Cross4All Platform" will appear, and the device is ready to be used.

Step 6: Settings. Data acquisition into patients' cloud-based PHR (Fig. 2b).

Step 7: Sharing. The patients grant permission to their PHR to selected doctor/s.

Step 8: Patient data utilization. The doctor logs in with their credentials (user-name/password) created via the web PHR or with PIN created during the first login. Once logged in, if the patients have given permissions to their file, the doctor can see a list of records of measurements. After selecting a patient, the doctor can "START RECORDING SESSION", and offers suitable treatment (Fig. 3a).

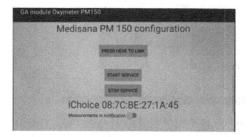
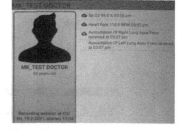

Fig. 2a. Communication between devices   **Fig. 2b.** Data acquisition into PHR

The workflow process in patients' mobile application contains the first seven steps (Fig. 3b). Before the first step, the patient must create their PHR on the project webPHR cloud system. The patient can access to webPHR through their mobile devices or info-kiosk in the public healthcare institution.

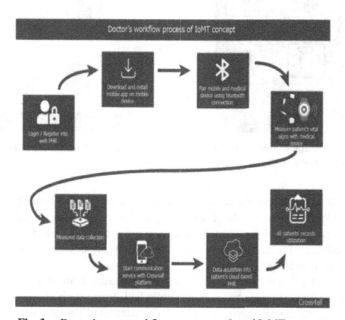

Fig. 3a. Doctor's apps workflow processes of used IoMT concept.

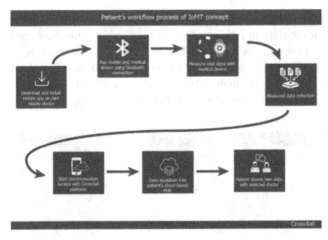

Fig. 3b. Patient's apps workflow process of used IoMT concept.

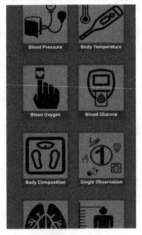

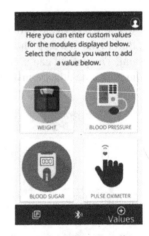

Fig. 4a. Mobile application for medical
staff – Observations

Fig. 4b. Modules for manually entering

The medical practitioners have an opportunity to enter values manually with mobile application Observations (Fig. 4a). For entering a value manually, the mobile application user first has to select option Observations and specific icon (Fig. 4b) and after that enter the measured value. In addition, the patients can also enter values manually with modules for manual data entering.

Medical professionals' medical devices, sensors, and applications are more complex, but they are still easy to use. An example would be Littman stethoscope - after pairing with the tablet, the mobile application has to be connected with the stethoscope. The medical professional selects a body area from the screen (Fig. 5) and starts the recording from the application or the stethoscope directly. For all recorded measurements to be synchronized, the mobile app needs some time (Fig. 6).

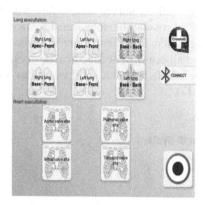

Fig. 5. Littman stethoscope application   **Fig. 6.** Data transfer to the app

All acquired data has been uploaded and can be viewed from WebPHR and shared with health care professionals (Fig. 7). It is a central point application for patient's PHR cross border that integrate all patient's healthcare data [2], including e-prescription and e-referral system (Fig. 8).

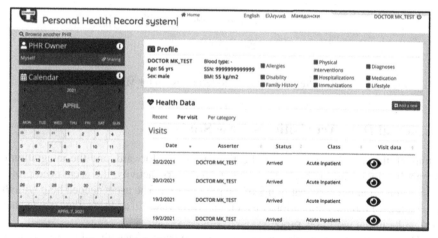

Fig. 7. Web PHR-shared data with a medical professional.

In the PHR, the patient's data are organized in Allergies, Disabilities, Family History, Surgeries, Immunizations, Diagnoses, Medication, and Lifestyle data of the patient. In addition, in the PHR, laboratory results and DICOM images can be displayed.

This Cross4all ecosystem integrated as applications and features, in full, provides a wide range of possibilities and capabilities [2]. Cross4all interconnected services contain the systemized cloud webPHR monitoring system for patients, cloud-based computing, data analysis and smart medical care, and all sensor-based medical devices, Bluetooth, Internet, mobile applications, cloud-based platform, etc. [13] as well as secure management of the data of each PHR [7, 16, 17].

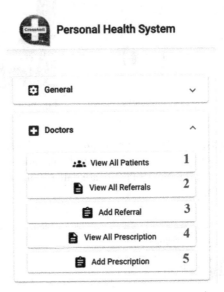

Fig. 8. Cross4all e-Prescription and e-Referral system for Medical professionals.

6   Actual Data Acquisition – Case Study

Although the installation of the applications is straightforward, and the integration of the applications with the backend does not need users' activity, the users need professional support at the very beginning of the pilot project. The users needed help with the installation of mobile applications on their mobile devices. In addition, they needed help during the initial workflow process.

The users were completely supported by a team from the Faculty of Information and Communication Technologies-Bitola, one of the project partners. Considering the user needs, they installed and prepared mobile devices with mobile applications, prepared web PHR manual, and mobile apps manual for both medical practitioners and patients. There were many trainings for users, mobile team, and doctors, some of them online. The eLearning platform with a digital health and healthcare literacy courses related to e-health, health information and services from electronic sources, and information how the patients and medical practitioners can effectively use the Cross4all tools (webPHRs, devices, mobile apps) was created.

The Pilot project implementation of Cross4all project was smoothly done because of the collaboration with all partners. Firstly, the level of digital healthcare literacy had to be increased for the patients and general practitioners who run the pilot. They created PHR for all project participants and studied the possibilities provided by webPHR and mobile applications with sensors. After that, the citizens were included: they were educated firstly, and then they created their PHR. The general practitioners provide sensors for citizens connected with webPHR with the Mobile application for patients for vital signs measurement for some distance citizens. Disabled people were also included in this project, and for this reason, the digital assets were accessible, providing WCAG compliance standards.

7 Conclusion

The more convenient methods of measuring and collecting medical data through wireless sensors and IoT devices have led to the possibility of having much more data for patient health tracking and medical analysis and self-care, especially needed in the pandemic. The Cross4all project intends to combine the power of cloud computing and service-oriented architectures and provide patient-centered care, including telecare possibilities, self-care elements and increasing evidence-based healthcare management in a cross-border environment where patients are mobile and not always physically available to their general practitioner. The health data collection and health data itself are the responsibility of the patient. However, since the collection is done by convenient medical devices, stored, and presented by a smartphone app, the patients are more likely to collect more data since they can always access it and see trends and some other insights regarding their health. One of the advantages of evidence-based medicine is that the end-users, i.e., patients, have to be aware of their current health situation and are willing to collect and measure more health data about themselves.

This approach has shown a need for greater attention from the authorities to increase the digital health educating of the health practitioners for this type of concept to be applied to a broader audience that is necessary for the future. In the next period, the number of available medical practitioners will not increase as the number of patients needing care and monitoring, especially for the patients with comorbidities that require constant monitoring of their vital signs.

This integrated Cross4all ecosystem of applications and characteristics provides a wide range of capabilities. Although the users initially feel slightly confused, the doctors and mobile teams become familiar with the workflow, mobile application, and PHR for a short time. The patients, especially the older people, have more significant user problems and had to increase their digital and e-health literacy.

This concept is necessary when the number of available medical staff decreases and the number of patients with chronic illness and comorbidities that require post-monitoring of their vital signs increases. Implementation of this concept needs much awareness of the benefits of increasing m-health and e-health digital health education and training, and additional staff support to get used to the number of patients.

As future work, we will try to capitalize the gained knowledge from the Cross4all project to integrate more countries in the system and to create the platform for multi-country healthcare collaboration. In addition, many other devices as IoMT sensors can be included in the application for measuring vital signs of life for citizens and professionals as well as their integration with backend PHR application.

Acknowledgment. Part of the work presented in this paper has been carried out in the framework of the project "Cross-border initiative for integrated health and social services promoting safe ageing, early prevention and independent living for all (Cross4all)", which is implemented in the context of the INTERREG IPA Cross Border Cooperation Programme CCI 2014 TC 16 I5CB 009 and co-funded by the European Union and national funds of the participating countries.

References

1. Savoska, S., Ristevski, B., Blazheska-Tabakovska, N., Jolevski, I.: Towards integration exposome data and personal health records in the age of IoT. ICT Innovations 2019, pp 237–246. Republic of Macedonia, Ohrid (2019)
2. Snezana, S., et al.: Cloud based personal health records data exchange in the age of IoT: the Cross4all project. In: Vesna, D., Ivica, D. (eds.) ICT Innovations 2020. CCIS, vol. 1316, pp. 28–41. Springer, Cham (2020). https://doi.org/10.1007/978-3-030-62098-1_3
3. Savoska, S., et al.: Design of cross border healthcare integrated system and its privacy and security issues. Comput. Commun. Eng. **13**(2), 58–63 (2019). (First Workshop on Information Security 2019, 9th Balkan Conference in Informatics (2019))
4. Maksimović, M., Vujović, V.: Internet of Things based e-health systems: ideas, expectations and concerns. In: Khan, S.U., Zomaya, A.Y., Abbas, A. (eds.) Handbook of Large-Scale Distributed Computing in Smart Healthcare, pp. 241–280. Springer International Publishing, Cham (2017). https://doi.org/10.1007/978-3-319-58280-1_10
5. Kadhim, K.T., Alsahlany, A.M., Wadi, S.M., Kadhum, H.T.: An Overview of Patient's Health Status Monitoring System Based on Internet of Things (IoT), Springer Science+Business Media, LLC, part of Springer Nature 2020, Wireless Personal Communications (2020)
6. Suganthi, M.V., Elavarasi, M.K., Jayachitra, M.J.: Tele-Health Monitoring System in a rural community through primary health center using Internet of Medical Things. Int. J. Pure Appl. Math. **119**(15) (2018)
7. Thilakarathne, N.N., Kagita, M.K., Gadekallu, T.R.: The role of the internet of things in health care: a systematic and comprehensive study. Int. J. Eng. Manage. Res. **10**(4) (2020)
8. Baker, S.B., Xiang, W., Atkinson, I.: Internet of Things for smart healthcare: technologies, challenges, and opportunities. IEEE Access **5**, 26521–26544 (2017)
9. Dang, L.M., Piran, M., Han, D., Min, K., Moon, H.: A survey on Internet of Things and cloud computing for healthcare. Electronics **8**(7), 768 (2019). https://doi.org/10.3390/electronics8070768
10. Qadri, Y.A., Nauman, A., Zikria, Y.B., Vasilakos, A.V., Kim, S.W.: The future of healthcare Internet of Things: a survey of emerging technologies, IEEE Commun. Surv. Tutor. **22**(2, Secondquarter 2020), 1121–1167 (2020)
11. Savoska, S., Jolevski, I.: Architectural model of e-health PHR to support the integrated cross-border services. In: ISGT Conference 2018, 16, 17 Nov 2018, Sofia (2018)
12. Mersini, P., Evangelos, S.: Personalized assistant apps in healthcare: a systematic review. In: 2019 10th International Conference on Information, Intelligence, Systems and Applications (IISA). IEEE (2019)
13. Tanaskoska, L., et al.: Optimizing bluetooth communication between medical devices and android. In: IX International Conference on Applied Internet and Information Technologies AIIT 2019. Zrenjanin, Serbia (2019)
14. Singh, R.P., Javaid, M., Haleem, A., Vaishya, R., Ali, S.: Internet of Medical Things (IoMT) for orthopaedic in COVID-19 pandemic: roles, challenges, and applications. Clin. Orthop. Trauma **11**(4), 713–717 (2020)

15. Blazheska-Tabakovska, N., Ristevski, B., Savoska, S., Bocevska, A.: Learning management systems as platforms for increasing the digital and health literacy. In: ICEBT 2019: Proceedings of the 2019 3rd International Conference on E-Education, E-Business and E-Technology, August 2019, pp. 33–37 (2019)

16. Hatzivasilis, G., Soultatos, O., Ioannidis, S., Verikoukis, C., Demetriou, G., Tsatsoulis, C.: Review of security and privacy for the Internet of Medical Things (IoMT). In: 15th International Conference on Distributed Computing in Sensor Systems (DCOSS) (2019)

17. Singh, G.: IoT for healthcare: system architectures, predictive analytics and future challenges. In: Singh, A., Mohan, A. (eds) Handbook of Multimedia Information Security: Techniques and Applications. pp. 753–773. Springer, Cham (2019). https://doi.org/10.1007/978-3-030-15887-3_36

Comparing Time and Frequency Domain Heart Rate Variability for Deep Learning-Based Glucose Detection

Ervin Shaqiri[1]([envelope]) [iD], Marjan Gusev[1,2] [iD], Lidija Poposka[2] [iD],
Marija Vavlukis[2] [iD], and Irfan Ahmeti[2]

[1] Innovation DOOEL, Skopje, North Macedonia
ervinshaqiri@gmail.com
[2] Ss. Cyril and Methodius University in Skopje, Skopje, North Macedonia
marjangusev@finki.ukim.mk

Abstract. Many researchers have been challenged by the usage of devices related to the Internet of Things in conjunction with machine learning to anticipate and diagnose health issues. Diabetes has always been a problem that society has struggled with. Due to the simplicity with which electrocardiograms can be captured and analysed, deep learning can be used to forecast a patient's instantaneous glucose levels.

Our solution is based on a unique method for calculating heart rate variability that involves segment identification, averaging, and concatenating the data to reveal better feature engineering results. Immediate plasma glucose levels are detected using short-term heart rate variability and applied a deep learning method based on Autokeras.

In this paper. we address a research question to compare the predictive capability of time and frequency domain features for instantaneous glucose values. The neural architectural search for the time domain approach provided the best results for the 15-min electrocardiogram measurements. Similarly, the Frequency Domain approach showed better results on the same time frame. Regarding the time domain the best results are as follows: RMSE (0.368), MSE (0.193), R^2 (0.513), and R^2 loss (0.541). The best results for the Frequency Domain approach the best results are as follows: RMSE (0.301), MSE (0.346), R^2 (0.45578), and R^2 loss (0.482).

Keywords: ECG · HRV · Deep learning · Glucose · Diabetes

1 Introduction

Diabetics Mellitus is a disease that affects people of all ages, and many people are unaware that their blood sugar levels are out of control. A variety of invasive, minimally invasive, and non-invasive approaches have been used to measure glucose levels [5] using different techniques and bio-chemical, physical or electrical-based scientific methods. Invasive approaches involve pricking your finger and evaluating the chemical properties of a blood drop, whereas minimally invasive procedures rely on a series of extremely small needles to evaluate

© Springer Nature Switzerland AG 2022
L. Antovski and G. Armenski (Eds.): ICT Innovations 2021, CCIS 1521, pp. 188–197, 2022.
https://doi.org/10.1007/978-3-031-04206-5_14

interstitial bodily fluid or other parameters. Electromechanical characteristics are used in non-invasive ways to scan the effects of elevated glucose levels on the skin, eyes, or other interstitial bodily fluid. Our solution is based on scanning the heart rate variability (HRV) with non-invasive electrocardiogram (ECG) sensors and determining a set of HRV factors processed by a new Deep Learning (DL) algorithm.

DL is a subset of Machine Learning (ML) utilizing artificial neural networks. The last decade in computer science has seen an explosion of application of DL methods, especially with the availability of cloud computing. Machines have enhanced their performance so much that today a person can run a DL process at home or even on a Raspberry PI [4].

The capabilities of DL solutions in other industries provide a good example of how to process biological data (in our case HRV) in order to predict instantaneous glucose levels. This study compares two DL solutions using different feature sets and keeping all input parameters and related variables the same. except for the features. In other words, one The first DL solution involves time domain HRV calculations, while the other solution involves frequency domain HRV calculations. The data and resources are provided under the umbrella of the Glyco project [7], conducted on a group of patients wearing a small apparatus that records the person's ECG signal and then relays that to a proprietary application that processes and annotates the appropriate information. These processed annotations comprise adequate information for calculating time and frequency domain HRV.

HRV is known as a good parameter for predicting different health conditions [12]. From previous research we have identified two methods for calculating HRV for a certain annotation file segmenting the signal into clean Normal-to-Normal intervals [13]:

- Averages - taking the average HRV for each segment.
- Combined - concatenating all segments into one long segment and calculating the HRV on that long segment.

In this research, we aim at predicting a quantity of a blood sugar level based on HRV measurements, and therefore it is a regression problem. Moreover, we are eager to find an answer to the following questions:

- RQ1. How do these different feature sets differ in predicting Glucose Levels?
- RQ2. Which HRV calculations comes closest to the correct levels?
- RQ3. Are Frequency Domain features better at predicting?

All these questions lead to the final research question:

- RQ4. Which HRV features are better at producing a solution for a noninvasive accurate prediction of Blood Sugar Level?

This article begins with a description of the related works in Sect. 2 and elaboration of methods in Sect. 3. Section 4 presents the results with their explanation, and Sect. 5 discusses and evaluates the results. Finally, the conclusions and future work are elaborated in Sect. 6.

2   Related Works

Kumar et al. [9] study the possibility of a mixture of time-frequency domain statistical features to classify the abnormalities in the electrocardiogram (ECG) signal. They use the public MIT-BIH Arrhythmia (MITDB) ECG database and Normal Sinus Rhythm Database (NSRDB). Utilizing decision trees the authors were able to get an accuracy up to 99.2% without addressing detection of glucose level. Habbu et al. [6] conducted a study on a specific Skillogies Pune India dataset with the goal of using ensemble learning to predict Cardiac Disease Among Smokers Based on HRV Features. Besides using the time and frequency domain HRV, they also use Nonlinear measurements as features for predictions. The validation of their approach shows accuracy of 95.20%, precision of 97.27%, sensitivity of 92.35%, specificity of 98.07%, F1 score of 0.95, AUC of 0.961, MCE of 0.0479, kappa statistics value of 0.9041, and MSE of 0.2189 for this specific dataset.

In a more recent study, Zhang et al. [14] analyzed polysomnography (PSG) data obtained from 2111 participants in the Sleep Heart Health Study. From which 1252 participants suffered CVD events (CVD group) and 859 participants remained CVD-free (non-CVD group). HRV measures, derived from time-domain and frequency-domain, were used at developing a predictive model with a final accuracy of 75.3%. It is worth noting that the authors do conclude that the frequency domain feature HF demonstrated to be an independent predictor of CVD outcome.

In another study Sankar et al. [11] explore the possibility of a non-invasive method of diagnosing Type 2 diabetes. The study consists of 75 subjects out of which 25 are non-diabetic. It is interesting that the authors use only non linear measurements from the HRV computations to do the predictions. They train an SVM with the features and are able to achieve an accuracy of 94.7%.

Rahman et al. [10] who aim to see the predictive capabilities of diabetes through different DL techniques: Convolutional Long Short-term Memory (Conv-LSTM), Convolutional Neural Network (CNN), Traditional LSTM (T-LSTM), and CNN-LS. The dataset contains records of 768 female patients aged at least 21 years among them 268 are diabetes positive and the rest are diabetes negative. The dataset has eight predictor variables like Pregnancies, Glucose, Blood Pressure, BMI, Skin Thickness, Insulin, Diabetes Pedigree Function, and Age for diagnostically predicting whether a patient has diabetes or not and one target variable named as "outcome". The authors conclude that the best model was generated by the Conv-LSTM with an accuracy of 97.26%.

DeepHeart [2] encompasses 14,011 users of a specific Apple Watch app. Each participant's data was then split into week-long chunks, and any weeks ≤ 30 min of continuous heart rate recordings were omitted this resulted in a total of 57,675 person-week data. The authors conclude that a multi-task long short-term memory (LSTM) yielded high accuracy results at detecting multiple medical conditions, including diabetes (0.8451), high cholesterol (0.7441), high blood pressure (0.8086), and sleep apnea (0.8298) [2].

Table 1. Distribution of age, weight, height, and BMI in the Glyco dataset [7]

Parameter	Male		Female		Total	
	Avg	Std	Avg	Std	Avg	Std
Age (years)	59.7	9.6	63.2	11.3	60.8	10.4
Weight (kg)	84.4	17.6	76.6	11.1	82.5	14.8
Height (cm)	161.9	46.5	164.1	4.7	171.3	8.6
BMI	26.1	8.6	28.5	4.1	28.3	4.5

As a summary of what is reported in the scientific community until year 2021 is that there are no related papers that show a successful method on detection of instantaneous blood glucose level from HRV. The research study based on DeepHeart [2] shows promising results of 84.1% using a proprietary dataset without revealing the feature engineering characteristics and dataset split.

3   Methods

3.1   Dataset

The Dataset consists of subjects that take part in the Glyco Study [7]. Although a total of 155 subjects participated in the study, the measurements of only 138 were with sufficient quality and pose both clean and continuous ECG measurements alongside manual Glucose measurement through finger pricking. It is also important to note that the subjects in this dataset are known to have health problems, more precisely stable coronary artery disease. The dataset contains 94 male and 44 female patients, where the distribution of age, weight, height, and BMI are presented in Table 1.

Our research focus is on the predictive capabilities between time and frequency domain features on short term recordings. We created six datasets of short term recordings with the following lengths: 30 s, 1 min, 5 min, 10 min, 15 min and 30 min recordings correspondingly labeled as D30S, D1M, D5M, D10M, D15M, D30M. The signals occur in the range of 1 h before and 1 h after measuring blood glucose through finger pricking.

A set of features is attached to each dataset along the target variable. A deep learning process is applied for each dataset, and the results are evaluated and compared.

3.2   Features

There are six time domain features and six frequency domain features. It is worth noting that calculating time domain features is more straightforward and requires less computation heavy operations thus are faster when compared to frequency domain features. The features and their description are presented in presented in Table 2, where by NN we denote the interval between successive

Table 2. Description of analyzed HRV parameters

Parameter	Description
SDNN	Standard Deviation of NN intervals
ASDNN	The Average Standard Deviation of NN intervals (minimum 5 min)
SDANN	The Standard Deviation of the averages of NN intervals (minimum 5 min)
NN50	Adjacent NN interval pairs differing more than 50 ms
pNN50	NN50 counts divided by total count of NN intervals
rMSSD	The square root of the mean of the sum of the squares of differences between adjacent NN intervals
ULF	Absolute power of the ultra-low-frequency band ($\leq$0.003 Hz)
VLF	Absolute power of the very-low-frequency band (0.0033–0.04 Hz)
LF	Absolute power of the low-frequency band (0.04–0.15 Hz)
HF	Absolute power of the high-frequency band (0.15–0.4 Hz)
LF/HF	Ratio of LF-to-HF power
Total power	The sum of all power bands
Distance	Total length of recording expressed in seconds
Timestamp	The time of the recording

normal heart beats. Note that SDNN, ASDNN, SDANN, NN50, pNN50, rMSSD are time domain HRV and ULF, VLF, LF, HF and LF/HF are frequency domain parameters.

An important aspect that requires attention is that some parameters require a minimum of 5-min recordings to be calculated. This means that the parameters ASDNN, SDANN, NN50, pNN50 are removed from the feature list for the D30S, D1M, D5M datasets.

3.3   Experiments

We have carried out a total of six experiments for each dataset and each feature set respectively, by assigning a single DL process to each dataset. In our experiments, we apply a deep learning neural architectural search. It is important to understand that a model trained on one time frame will and should not be used in predicting on another time frame.

The experiments (DL processes) are conducted with AutoKeras [8], as an automated ML system based on Keras [3], which is just a wrapper of Tensorflow DL framework [1]. The goal of the experiments is to predict the estimated instantaneous blood glucose level from different HRV features which are in turn obtained from an ECG signal.

The outcome of the conducted experiments is a predicted value and not a class, which defines it as a regression problem. Autokeras handles this issue by just supplying the datasets and providing values for the following parameters:

– the structured data regression class
– epochs - number of epochs for each trial

- max number of trials - the number of different architectures to try
- batch size
- metrics - which metrics to report while training
- objective - which metric to track and develop upcoming architectures according to the metric maximization or minimisation
- callbacks - custom callback which the user wants to use, the same as manual Keras callbacks (optional)

3.4   Evaluation Metrics

To evaluate a regression problem, we use the following set of metrics.

Mean Squared Error (MSE) measures the average of the squares of the errors and is calculated by (1) as an average squared difference between the estimated value and the actual value.

$$MSE = \frac{1}{n}\Sigma_{i=1}^n \left(\frac{d_i - f_i}{\sigma_i}\right)^2 \tag{1}$$

Root Mean Squared Error (RMSE) is calculated by (2) as an absolute measure of the goodness for the fit. In other words the differences between value predicted by a model or an estimator and the value observed.

$$RMSE = \sqrt{\frac{1}{n}\Sigma_{i=1}^n \left(\frac{d_i - f_i}{\sigma_i}\right)^2} \tag{2}$$

R squared (R^2), calculated by (3), is the proportion of the variance in the dependent variable that is predictable from the independent variable and usually is expressed as a percentage value. Sometimes it is called the coefficient of determination. as a measure of how close the data is to the fitted regression line (usually expressed as a percentage). Having an R^2 value trending towards 100% shows that the regression line is a good fit for the data.

$$R^2 = 1 - \frac{RSS}{TSS} \tag{3}$$

R squared loss (R^2 loss) is the numerical representation of how bad a model is predicting. The more closer to 0 the more accurate the model is at predicting the given data.

4   Results

Each experiment involved the same script where in the case of measurement less than five minutes some time domain features were dropped as discussed previously. The goal of each experiment was to minimize MSE as much as possible. Close attention was also paid to avoid data overfitting. Each experiment took roughly 24 hours to complete. Additionally, R^2 is used to evaluate how well an architecture generalized on data compared to other architectures.

Table 3. Performance of DL models for each dataset trained on time and frequency domain features

Dataset	Time domain features				Frequency domain features			
	MSE	RMSE	R^2 score	R^2 loss	MSE	RMSE	R^2 score	R^2 loss
D30S	3,978	1,267	0,47710	48,710	0,114	0,441	0,32775	34,030
D1M	1,498	1,024	0,17341	18,341	0,224	0,088	0,10483	11,887
D5M	0,206	0,385	0,12334	15,538	0,672	0,659	0,16458	15,751
D10M	0,162	0,314	0,10080	11,080	0,981	0,777	0,02335	01,994
D15M	0,193	0,368	0,51280	54,513	0,346	0,301	0,45578	48,187
D30M	0,090	0,232	0,23181	24,183	0,111	0,246	0,19894	20,771

A description of the architectures of each best model trained on time and frequency domain HRV features are presented in Table 4. Additionally, Table 5 presents the best architecture choice by AutoKeras when trained on Frequency Domain features of HRV.

A drawback using this automated DL process is that it is a black box solution. In other words we do not know the rationale behind which Autokeras decides to change the underlying architecture. The results for each best model generated from AutoKeras tested on the test set after training are presented in Table 3. The best performing model is the one generated for D15M with an R^2 value of 51 for the Time Domain approach, while for the Frequency Domain approach the best model is also on the D15M with an R^2 value of 46.

5 Discussion

Comparing the different architectures from automated DL process, we notice different architectures involved in each dataset. However, when it comes to certain hyperparameters we observe that the models share the same Optimizer, Learning Rate, Decay, Loss, Activation function and dropout layers which can be seen in Table 4 and Table 5. In other words the best performing optimizer came out to be Adam at a learning rate of 0.001 with no decay. Regarding the loss function the best performing function was shown to be MSEL, while the best activation function function ReLu.

The main difference between each architectures in the number of dense layers alongside the number of neurons for each layer. Interestingly some architectures share the same number of dense layers but greatly differ in the number of neurons in each layer. For example, we notice that D30S, D1M, D5M, and D15M perform best on architectures with three dense layers. However, D30S has 16,24 and 32 neurons per layer while D15M has 512,512, and 128 neurons respectively. The same goes for the D10M and D30M, where each perform best on two dense layers where D10M has 32 and 32 neurons per layer and on the other hand D30M has 512 and 256 neurons per layer.

Table 4. Best model architectures for time domain approach

Dataset	Input layer	Dense layers	Number of neurons	Batch layers
D30S	Shape (0, 6)	3	16, 24, 32	3
D1M	Shape (0, 6)	3	32, 32, 16	0
D5M	Shape (0, 6)	3	64, 16, 64	0
D10M	Shape (0, 15)	2	32, 32	2
D15M	Shape (0, 15)	3	512, 512, 128	2
D30M	Shape (0, 15)	2	512, 256	2

MSEL = Mean Squared Error Loss

Table 5. Best model architectures for frequency domain approach

Dataset	Input layer	Dense layers	Number of neurons	Batch layers
D30S	Shape (0, 6)	4	32, 32, 32, 32	1
D1M	Shape (0, 6)	3	32, 32, 256	0
D5M	Shape (0, 6)	3	128, 128, 128	2
D10M	Shape (0, 6)	2	32, 26	0
D15M	Shape (0, 6)	5	64, 128, 128, 128, 32	0
D30M	Shape (0, 6)	1	128, 128	0

Interestingly when we compare the architecture choices for the Frequency Domain approach we observe that when it comes to the Optimizer, Learing Rate, Decay, Loss, and Activation Functions the values are the same with the architectures generated by the Time Domain approach. However, the architecture choices differ quite a lot. For example, the number of hidden layers is approximately the same 4,3,3,2,5, and 1 for the D30S, D1M, D10M, D15M, D30M. But the number of neurons per layer changes a lot. For example, a range of neurons from 16–512 is found for the Time Domain architectures, a range of neurons from 32–128 for the Frequency domain architectures.

It is interesting to notice how the datasets that have longer time spans have better predictive capabilities. We believe this is due to the fact that longer time spans allowed for the other features to be present in the dataset, in other words, the features that required a minimum of 5-min recordings are not present in the 30 s, 1 m, and 5 m datasets which could be the reason that the rest of the datasets are better and generating models for prediction.

The best achieved result is the D15M dataset where the R^2 is 0,51280. The value is not ideal but the results can be improved. In order to make improvements more data would help out, but also a longer training session would yield better results. We also observe that the second-best result is obtained on the HRV calculations from the D30S dataset with an R^2 of 0,47710. The results of both datasets make for a good case on whether a combination of both datasets could yield a model that would better generalize the data?

Regarding RQ1, we conclude that the automated deep learning process differed when deciding on the architecture where the number of hidden layers and especially the number of neurons varied quite a lot. This is due to the fact that additional features in the Time Domain parameters seems to have pushed the Automated Deep Learning to go for a more shallow architecture but increase the number of neurons.

In relation to RQ2 analyzing the different metrics shows that both methods are roughly the same at giving the value of the sugar level.

With reference to RQ3 frequency domain features are not better at predicting Blood Sugar Level. Ultimately this leads to answering the final Research Question where the best HRV features for predicting noninvasive accurate Blood Sugar Level are the Time Domain Features. This is evident when comparing R^2 values, where for example for D30S we have a value of 0.477 for Time Domain, wile for the Frequency Domain on the same experiment we have a value of 0.328.

Addressing the RQ4, we conclude that ASDNN and SDANN are the best Time Domain parameters for our models, and LF for Frequency Domain parameters.

6 Conclusion

In a nutshell, we have shown an implementation of Autokeras on training conduct from two Domains of HRV. Even though these two domains differ greatly in the way they are calculated, where the Time Domain is a statistical calculation and the Frequency Domain is a spectral calculation we do see some similarities in the architectures such as the loss function being the mean squared error, the Adam optimizer, the learning rate 0.001 and the activation function being ReLu for each dataset of each feature set. We also notice the number of hidden layers for the Time Domain approach does not exceed 3 while for the Frequency Domain we have models with 4 and 5 layers. Interestingly, the number of neurons differ quite a lot where the Time Domain models have a range of 16–512 while the Frequency domain have a range of 32–128.

Furthermore, the best architecture trained on time Domain features consists of three layers with 512,512, and 128 neurons, Adam optimizer, a learning rate of 0.001, loss of MSE, and two Batch Normalization Layers. While the best architecture trained on Frequency Domain features consists of five hidden layers with 64,128,128,128, and 32 neurons, Adam optimizer alongside a learning rate of 0.001, and loss function of MSE. Where the R^2 results for the Time Domain model is 51.280 while for the Frequency Domain model R^2 is 45.578. According to the results the authors conclude that the Time Domain features is better at identifying estimated instantaneous blood glucose level.

To conclude, the results show that the different feature sets made the automated deep learning process generate different architectures for fitting the data. Additionally, both features somewhat came close to the same prediction value on the test sets. However, Frequency Domain Features showed sub optimal prediction compared to the Time Domain Feature set. The authors conclude that the Time Domain Features are better predictors for a noninvasive accurate prediction of Blood Glucose Level.

References

1. Abadi, M., et al.: TensorFlow: a system for large-scale machine learning. In: 12th USENIX Symposium on Operating Systems Design and Implementation (OSDI 2016), pp. 265–283 (2016)
2. Ballinger, B., et al.: Deepheart: semi-supervised sequence learning for cardiovascular risk prediction. In: Proceedings of the AAAI Conference on Artificial Intelligence, vol. 32 (2018)
3. Chollet, F., et al.: Keras: The Python Deep Learning Library. Astrophysics Source Code Library pp. ascl-1806 (2018)
4. Dürr, O., Pauchard, Y., Browarnik, D., Axthelm, R., Loeser, M.: Deep learning on a raspberry pi for real time face recognition. In: Eurographics (Posters), pp. 11–12 (2015)
5. Gusev, M., et al.: Noninvasive glucose measurement using machine learning and neural network methods and correlation with heart rate variability. J. Sens. **2020** (2020)
6. Habbu, S.K., Dale, M.P., Ghongade, R.B., Joshi, S.S.: Comparison of noninvasive blood glucose estimation using various regression models. In: Thampi, S.M., et al. (eds.) SIRS 2019. CCIS, vol. 1209, pp. 306–318. Springer, Singapore (2020). https://doi.org/10.1007/978-981-15-4828-4_25
7. Innovation Dooel: a doctor in your pocket measure ECG & glucose levels with a small, non-invasive, wearable monitor (2018). http://glyco.innovation.com.mk/
8. Jin, H., Song, Q., Hu, X.: Auto-keras: an efficient neural architecture search system. In: Proceedings of the 25th ACM SIGKDD International Conference on Knowledge Discovery & Data Mining, pp. 1946–1956 (2019)
9. Kumar, R., Verma, A.R., Panda, M.K., Kumar, P.: Hrv signal feature estimation and classification for healthcare system based on machine learning. In: International Conference on Machine Learning, Image Processing, Network Security and Data Sciences, pp. 437–448. Springer, Singapore (2020). https://doi.org/10.1007/978-981-15-6315-7
10. Rahman, M., Islam, D., Mukti, R.J., Saha, I.: A deep learning approach based on convolutional ISTM for detecting diabetes. Comput. Biol. Chem. **88**, 107329 (2020)
11. Sankar, P., Cyriac, M.: A non-invasive approach for the diagnosis of type 2 diabetes using HRV parameters. Int. J. Biomed. Eng. Technol. **26**(1), 71–83 (2018)
12. Shaqiri, E., Gusev, M.: Deep learning method to estimate glucose level from heart rate variability. In: 2020 28th Telecommunications Forum (TELFOR), pp. 1–4. IEEE (2020)
13. Shaqiri, E., Gusev, M., Poposka, L., Vavlukis, M.: Correlating heart rate variability to glucose levels. In: International Conference on ICT Innovations, web proceedings, pp. 1–10 (2020)
14. Zhang, L., Wu, H., Zhang, X., Wei, X., Hou, F., Ma, Y.: Sleep heart rate variability assists the automatic prediction of long-term cardiovascular outcomes. Sleep Med. **67**, 217–224 (2020)

Digital Shift: Assessment of Mental States Through Passive Mobile Sensing

Evgenija Krajchevska[1]([✉]), Nina Petreska[1]([✉]), Ognen Handjiski[1]([✉]),
Sandra Andovska[1]([✉]), Bojan Ilijoski[2]([✉]), Petre Lameski[2]([✉]),
Panche Ribarski[2]([✉]), and Biljana Tojtovska[2]([✉])

[1] LOKA, 350 2nd Street, Suite 8, Los Altos, CA 94022, USA
{evgenija,nina,ognen,sandra.andovska}@loka.com
[2] Ss. Cyril and Methodius University, Skopje, North Macedonia
{bojan.ilijoski,petre.lameski,pance.ribarski,
biljana.tojtovska}@finki.ukim.mk

Abstract. Stress remains a prevalent risk factor associated with various diseases. As humanity enters a stage of enforced confinement as a response to the recent pandemic, cases of anxiety, stress, insomnia and depression begin to skyrocket thus, exposing a darker social scenery that challenges our well-being and overall mental state. Motivated by the recent research in the area of mental health assessment, we investigate ways of using ensemble learning methods applied to mobile sensing data for estimating user's stress levels. A growing number of sensors embedded in smartphones have resulted in the birth of the mobile sensing research area while at the same time providing plenty of possibilities for mobile applications designed for helping and changing the way of our living. In addition to stress levels prediction, tracking of mood is also included by the utilization of an emotion classifier. Both solutions are incorporated as functionalities of a mobile application which we describe in greater detail further. Accurate and efficient assessment of stress levels in a timely manner could greatly help the administration of proper measures for mental health improvement.

Keywords: Passive sensing · Stress · Emotion recognition · Ensemble methods · Convolutional Neural Networks

1 Introduction

With over 3 billion users globally, mobile phones are now the most omnipresent consumer device in the world. Unfolding the course of day-to-day living, they have resulted in a massive increase in personal comfort and effectiveness. Now more than ever, mobile phones have on-board sensors needed to collect data about their owner's behavior. Being equipped with powerful sensing and computation capabilities, they collect these data unobtrusively which in turn allows for accumulating people's social experiences such as speaking rates in conversation

L. Antovski and G. Armenski (Eds.): ICT Innovations 2021, CCIS 1521, pp. 198–220, 2022.
https://doi.org/10.1007/978-3-031-04206-5_15

and who we interact with, everyday activities including physical activity and sleep, as well as mobility patterns (e.g. time spent at any location traceable). There is growing interest, especially in psychological and medical investigation, in using mobile phones for behavioral observation [1–5]. Although they are not the only mobile-sensing devices that collect ecologically validated behavioral information, they do succeed in addressing methodological challenges facing the field as it strives to become a truly behavioral science.

Having great potential in health-related research, statistical models exploiting sensor-derived features promise to advance human-computer interaction and building the applications to use them is currently being explored. With the StudentLife study [6], alterations in students' mental and emotional health are captured by linking behavioral patterns that manifested over long periods to aspects of traditional health-related surveys. The dataset has allowed researchers to conduct proof-of-concept studies to investigate the relationship between passive sensing measures and various mental health outcomes (namely depression, loneliness, stress) [7–9]. Despite the focus being on a limited number of sensing-derived features, these studies have demonstrated the usefulness of mobile phone sensing applications in obtaining unbiased data about users and their environment.

What we present in this paper is "Emphasize", a mobile phone application able to effectively measure users' stress relying on passively collected data and minimal human-interference, including an option to upload a photograph of themselves to additionally gain insights on their emotional well-being. To understand and diminish stress we have to be able to recognize it and already existing approaches comprise of assessing saliva cortisol levels [10], administering questionnaires [10] and several others which truly are invasive, expensive as well as infuriating. That is why, we aim to model stress level dynamics by lessening reliance on self-reported methods subject to bias and by administering an intuitive, user-friendly application we hope to further deepen sensing-derived research. To accomplish this, our study addresses several aspects:

1. Exploration of StudentLife study participants to gain insights on meaningful patterns in relation to stress and applying clustering techniques.
2. Processing methodology of sensing data.
3. Building predictive models to capture subjects' variety of emotions, namely stress, anger, happiness and such.
4. Development and deployment of the mobile phone application "Emphasize".

The rest of this paper is organized as follows: Recent related works of passive sensing exploitation and stress detection using current technologies are stated in Sect. 2. Sections 3 and 4 focus on investigating already available data and the processing pipeline. Section 5 discusses the learning framework, while results are reported in Sect. 5. Development of the applications and server architecture is described in Sect. 6. Finally, concluding remarks are given in Sect. 7.

2   Related Work

The development of mobile sensing applications from the ground up is a process that is usually very time-consuming, and requires human resources as well as technological expertise. Since the early 2000s, human-computer interaction (HCI) researchers have been dwelling into sensor-based modelling of human behavior, with availability and interruptibility being common topics [11–14]. Ever since, the transitions from studying these models to deploying systems based on such models introduced new challenges and having become an increasingly used technique to obtain experimental data, mobile sensing is now considered a research area on its own.

Several publications have shown that mobile phones are an appropriate instrument for acquiring pertinent data that can be used to assess numerous aspects of human behavior and classify different mental states [15–18]. The StudentLife study, which we exploit in our research and application, investigates this relationship in Dartmouth students during an academic term. Using the same dataset, Harari et al. analyzes the changes of students' activity and sociability behaviors over a term via the accelerometer and microphone sensors [19]. With regards to quantifying stress in a unobtrusive and non-burdensome way, several studies have presented alternatives including mobile phones and wearable devices. Current methods mainly rely on physiological signals, for instance, blood pressure, body temperature and respiration [20]. Poh et al. provides us with evidence that a wearable device is a feasible alternative for measuring electrodermal activity (EDA) over a period of time [21]. Although accurate, these methods present a limitation that they need to be carried at all times to allow for continuous monitoring. "StressSense" tries to alleviate these difficulties by detecting stress based on speech analysis and variation of articulation [22]. It is tested in several environments to be capable of adapting to different scenarios. Nevertheless, in a real-life scenario, this method may lead to speech misinterpretation, and consequently of emotions.

In another study, Sano and Picard make use of a wrist sensor, mobile phone usage and surveys to find markers, both physiological and behavioral, for stress [23]. Findings show that passively collected data includes features relating to stress level dynamics, in particular, higher stress levels were shown to correlate with fewer text messages sent as well as certain screen on/off patterns. More accordant to the data we use, Halm et al. introduce the "Mobile Photographic Stress Meter" [8]. Selecting one of sixteen images that describes their stress levels, results are validated using the Perceived Stress Scale (PSS) questionnaire answered [24]. Results show that the application's score strongly correlates with PSS scores, demonstrating its effectiveness to measure stress.

As in previous research, what we propose is a framework to learn relevant features of passively collected data and make stress predictions using them. Moreover, if a user is willing to capture a self-portrait, they will be assessed of additional seven mental states. We exploit supervised learning methods which use labeled data to classify the outcomes accurately, both for the stress and the optional emotions predictions. To do so,

Relying on this type of data we by reducing reliance on self-reported methods that are susceptible to bias.

Regardless, an expert-based approach that utilizes knowledge of psychological scientists would be the best way of handling this delicate matter.

3 Data Investigation

In absence of smartphone sensing data needed for predicting one's stress levels, we use the StudentLife dataset as our starting point. The StudentLife study consists of passive and automatic sensing data of 48 Dartmouth students over a 10-week term [6]. It includes over 53 GB of continuous data, over 30.000 self-reports and pre-post surveys. The continuous sensing system comprises ten sensors and is coupled with behavioral classifiers that run 24/7 with duty cycling. Physical activity inferences are obtained through the usage of the accelerometer, while audio inferences are based on microphone recordings and include conversation detection. Scans from GPS, Bluetooth, and Wi-Fi aid in detecting location. Other sensors report prolonged periods longer than 1 h, covering phone lock, phone charge and the light sensor that captures ambient light.

A subset of the available dataset is used, namely, physical inferences, GPS data, light sensor, phone lock and phone charge. How these sensors are implemented in our application and how the inference differs from the ones used in the dataset is explained in more detail in Sect. 4, whereas the exploratory analysis is portrayed in the following subsections.

3.1 EMA Questionnaries

To gain insight into students' well being, the StudentLife smartphone application integrates surveys through an ecological momentary assessment (EMA) component to probe students' states (e.g., stress, mood) across the term [30]. A number of well known pre-post health and behavioral surveys were administered at the start and at the end of term. Our aim was to examine possible correlations of the responses from the students.

Grouping the responses allowed us to see which categories students responded to the most, the largest belonging to the stress questionnaire. Following is sleep, which we consider to be closely linked with stress. Our main interest being stress predictions, we explore how the responses from the other questionnaires correlate with it. A user answers a questionnaire with an ordinal variable describing the agreeableness with the statement and is available to do so multiple times a day. Such responses are averaged per user.

Using Pearson coefficient [31], we observe moderate correlations with the strongest being in the questionnaires regarding student's mood, where their happiness levels have a positive correlation of 0.44 while the rating of their sadness has a negative correlation of -0.53. Stress levels can be assessed with a better accuracy, however, the feedback to this survey is unsatisfactory and impossible to work with given the fact it has only 267 responses.

As a final step, sensing data is consolidated with the responses of stress surveys where users rate their stress levels on a scale from one to five, allowing further inspection of behavioral patterns collected leading to a particular answer.

3.2   Sleep Inference

Aforementioned, we hypothesized that sleep is closely related with stress. The causal relationship between quality of sleep and stress has been studied extensively and a myriad of factors were shown to impact the sleep response to stress, such as event appraisal [32], stressor chronicity and substance misuse [33], coping behaviour [32,34], cognitive-emotional and psychological defects [35]. Furthermore, individuals vary greatly in the degree to which they experience stress-related sleep problems, thus, even mild challenges, including unfamiliar sleep environments, small circadian shifts and ill-timed low-dose stimulants, produce transient sleep difficulties in many individuals [36]. Therefore, accurately predicting sleep duration may prove beneficial in the overall assessment of stress levels.

Using mobile sensors in sleep duration prediction is a rather novel approach with many applications for sleep monitoring emerging during the last couple of years. A study which uses a subset of the sensors included in the StudentLife dataset can classify whether a 10 min segment was reported as sleep with average accuracy of 88.8%. Furthermore, the same approach (EMA) for reporting data about sleep times was used [37]. The combined usage of a vast number of smartphone sensors can predict sleep duration accurately, however we want to inspect a more subtle approach without relying on activity recognition and location services, thus omitting the use of sensitive data. A recent study successfully develops an algorithm by analyzing users' interaction with their smartphones, in particular, only the smartphone screen interaction data (screen ON and OFF) [38]. Following their example, we use light, phone lock and phone charge from the StudentLife dataset to produce a relevant sleep feature. The light sensor refers to ambient light and captures the time spent in a dark environment, while lock and charge measure the duration for which the phone was locked or charging. Furthermore, there are 1390 answers to the sleep survey where the amount of sleep in hours is provided, therefore we can couple them with the sensing data of its respondents.

Several methods are explored and evaluated. Predicting sleep duration based on a single sensor yields poor results, with mean errors as large as 1.9 h for phone charge. Slightly better predictions are achieved with phone lock and light sensors, with an error of 1.8 h, using Support Vector Regression with a polynomial kernel. During investigation, moderate amount of data discrepancies were discovered. For example, maximum daily duration from all three of the sensors being 2.1 hs, while the average of responses the user submitted being 15 h of sleep did not correspond. These, as well as "0 h of sleep" responses were removed. Moreover, we experiment with a derived feature by taking either the mean of the longest duration from the sensors for each day in one case, or the maximum of the three in the other. The latter approach yielded better results, whereas working with the mean resulted in larger errors among users with high variability sensing data, with a mean error of 1.5 h.

Finally, the best result is achieved by looking only at the maximum daily duration from the sensing data. The distributions of the testing sets are shown in Fig. 1.

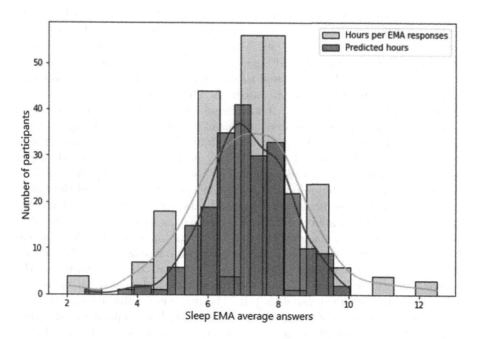

Fig. 1. Comparison of sleep duration on testing data.

As noted by many other studies [37], missing sensor data and misreports have an impact on the overall accuracy and performance of the models. Despite having limited data to work with, we show that only a handful of sensors residing in every smartphone are more than enough to infer duration of sleep.

3.3 Clustering Experimentation

One of the ideas was to build a model to predict stress levels for the user using data from the users with similar behavior i.e., similar-user model. In order to find users with similar behavior, multiple approaches were tried. The best outcomes came from one of the simplest methods. In particular, the vector used for the user was generated by counting the values above the mean value from all instances from all users for each feature i.e. the threshold value for each feature separately. Many tests were conducted in order to determine the attributes that should be used in the vector for user profiling. The vector that was used for clustering the users contained count features for the number of days they were stressed, neutral or not stressed and number of days they were running, waking, stayed in dark and kept their phone locked more than the mean threshold value.

Using the most representative days from one user, measuring statistical metrics from both days, and creating a similarity metric where the days of each pair of users are compared and abridged into one value were the other approaches we attempted.

Cluster analysis is a promising method for identifying and describing subgroups of individual cases defined by similarities among multiple dimensions of interest (e.g., coping strategies) [39]. The k-means algorithm requires to specify the number k of clusters. To find a good approximation of k the elbow method was used, which is a heuristic used to determine the numbers of clusters in a dataset and can be traced back to Robert L. Thorndike [40].

The number of clusters $k = 4$ was chosen as the number of final clusters. Visual inspection of cluster centers and mean scores on the clustering variables suggested very similar patterns across the four clusters. The first cluster represented individuals who had perceived very positive days, the second cluster represented individuals who were stressed for the majority of the time, the third cluster represented individuals who had perceived neutral days and the fourth cluster represented a combination of users who did not fit into the previous clusters. Therefore, the number of stressed days was the most significant factor in placing users into their clusters. This is a flaw in this approach, which is the reason it is very ineffective in case a user is placed in a wrong cluster. The second flaw was that the clusters were imbalanced. To evaluate the clusters, we use the Kappa Statistic or Cohen's Kappa which is a statistical measure of inter-rater reliability for categorical variables [41]. The maximum value means complete agreement; zero or lower means chance agreement. Each separate cluster produced outstanding results in its own cluster, with a Kappa statistic value in range between 0.47 and 0.82, and accuracy ranging from 0.64 to 0.78. However, when testing with users from other clusters the Kappa statistics drop to 0.3, due to the fact that each cluster has a certain day that it favors, i.e., the mode in the cluster.

Since the application lacks prior knowledge for the potential new user, there is no accurate way to put the user into a cluster. Hence, this approach could be better suited to larger datasets and until the user's data has been gathered for some period.

4 Data Processing

4.1 Inferring Stress

Model training and evaluation for the main part of our mobile phone application, predicting stress, is conducted using the StudentLife study consisting of passive and automatic sensing data of 48 Dartmouth students over a 10-week term [6]. Although popular approaches utilizing this dataset work with all of its extracted features, we focus on the following:

- **Objective-sensing data:** physical activity (stationarity, walking, running as well as unknown data). We use the percentage of each activity conducted daily as features.
- **Location-based data:** GPS coordinates, from which we infer daily minimum, maximum and average temperature using Meteostat, a free online service which provides weather and climate statistics, integrating it with a Python API [42].
- **Other phone data:** light, screen lock/unlock and phone charge. Duration and frequency of usage were used.
- **Self-reports:** sleep and stress, the latter used as a target variable.

Additionally, depending on the date, we added a feature which tells whether the date is a weekday or a weekend. Classifiers used in the StudentLife study run 24/7 with duty cycling, some of which make inferences continuously for one minute and generate a classification every two-three seconds (for example, physical activity) while others collect data every ten minutes (as GPS coordinates are). Phone data is recorded when the activity is registered for a significant long time i.e. more than one hour. To be able to take advantage of these findings, data preprocessing was necessary. The functions and techniques used are listed below.

1. **Data aggregation:** Since our idea was to make predictions every day, we decided to aggregate the available data on a daily basis. Objective-sensing data was used as the percentage of activity belonging to each class (including unknown activity), GPS coordinates aided in acquiring the temperature a subject was exposed to during the day, and other phone data was calculated as the mean of time recorded for each. Self-reports, where available, were used either as the mean value (if answered multiple times a day) or as value the user entered. Before we settled on daily aggregated data, we experimented with the days divided in two, three and four parts. Since results did not differ as much, there was no use of wasting computation power. The target variable for our classifier is stress, which ranges from 1 (being very stressed) to 5 (feeling very calm).

2. **Handling missing values:** The way we aggregated the data gave us crucial information, namely missing values were ranging from 12–33% per user. Absence of data reduces statistical power and may bias the estimation of parameters in the model. This also questioned the representativeness of the samples. To deal with this, a K-Nearest Neighbor imputation was conducted, per user, following a simple rule-of-thumb practice where the number of neighbors chosen equaled the square root of available data for a user. Several other methods were also tried out, including filling missing values with the feature's mean or median, linear and time methods of interpolation, multivariate feature imputation (using both mean and median), the MICE algorithm and such, but did not yield satisfactory results [41]. We would like to note that the stress variable which we used as a target for the classifier was imputed as well.

3. **Sampling:** When joining the data from users, another challenge we faced was the high disbalance in class distribution. This would have led to the machine learning algorithm to learn that minority classes were not as important as majority classes. To overcome this, we applied random under-sampling of the dataset, synthetic minority oversampling technique (SMOTE) [25] and a mixture of both to arrive at the median class label - neutral, denoted as "Over/Under-sampling". Additionally, we performed the same techniques on three class labels where levels 1 and 2 were combined to denote feelings of stress, 3 was left to be neutral, and 4 and 5 represent a feeling of calmness.

After processing the signals and features was done, the subsequent step was to feed them to several machine learning models and choose the best among them. We choose a decision tree as a baseline model, which is a method for approximating discrete-valued target functions, in which the learned function is represented by a decision tree. Learned trees can also be transformed to sets of if-then rules to improve human readability [26]. The objective of a decision tree is to specify a model that predicts the class value given that some input information is provided. In the study we put more focus on ensemble methods, which use multiple models to obtain better predictive performance that may or may not be obtained from any single model. More on this in the following section.

4.2 Emotion Recognition

The other feature of "Emphasize" is its ability to provide users with insights on their well-being, by showing statistics on several emotions present on the self-portrait photograph they may choose to upload. These emotions which are shown in the facial expressions fall in the following categories: anger, disgust, fear, happiness, sadness, surprise, and neutral feelings. To accomplish this, we take advantage of FER-2013, The Facial Expression Recognition 2013 dataset comprising around 35.000 48 × 48 pixel grayscale images of faces [27]. Faces in the dataset have been registered so that the face is centered and occupies approximately the same amount of space in each image. Prior to model training, images undergo several modifications:

1. **Rescaling:** To transform every pixel value from range [0, 255] to [0, 1], rescaling is necessary. Scaling the images in this way ensures that each input parameter (in this case, pixel) has a similar data distribution. Without it, high pixel range images will have large amounts of votes to determine how to update the weights during learning and the classifier will fail to generalize well.
2. **Random zoom:** We augment the photographs such that during training, a picture will be randomly zoomed into or zoomed out of done in the range [0.7, 1.7].
3. **Random flip:** We apply a horizontal flip to the data to add a mirror effect in order to increase generalization capabilities of the model we apply.

Considering the model, a deep learning architecture has been applied, highlighting the few important concepts below:

- **Convolutional Layers:** The mathematical operation of convolution applies a set of locally shared filters (or kernels) to obtain the most representative features. Convolutional layers convolve the input and pass its result to the next layer. These layers are suitable for image processing in contrast to fully connected (dense) layers because they reduce the number of free parameters.
- **Batch Normalization Layers:** To stabilize the learning process and reduce the number of training epochs, batch normalization layers aid the optimization process by "normalizing" the data inside the neural network. These layers reduce the internal covariate shift and diminish the reliance of gradients on the scale of the parameters or their underlying values.
- **Pooling Layers:** Convolutional networks often include local or global pooling layers to streamline the underlying computation. Pooling layers reduce the dimensions of the data by combining the outputs of neuron clusters at one layer into a single neuron in the next layer. This further reduces the computational cost by reducing learnable parameters and it provides translation invariance to the internal representation.
- **Dropout Layers:** These help correct overfitting, caused by the network learning spurious patterns in the training data. They create a kind of ensemble network, where a fraction of the layer's input units is dropped every step of training, making it harder for the network to learn the aforementioned patterns.

5   Learning Framework and Results

In this paper, we aim to present evidence that mental states, such as stress, can be efficiently predicted using a fast, distributed gradient boosting method, namely Light GBM. To do so, we make use of the StudentLife dataset [6]. Capturing the wider range of emotions utilizes the FER-2013 dataset [27]. As stated, the latter feature of "Emphasize" is optional, therefore two separate classifiers are trained.

5.1   Stress Classifier

In the previous section, we stated that we concentrated on ensemble methods for stress predictions. The choice for this stems from two observations: firstly, to our knowledge, excluding random forests, there is no instance of using the methods we experiment with in predicting stress working with the StudentLife data [6] and secondly, they are known for their impeccability to improve predictive performance. Mentioned previously in Sect. 3, we conducted a K-Nearest Neighbor imputation to handle missing values. We worked both with data where the target variable "stress" was imputed and was not. Within each dataset, several variations were tested:

- Five "stress" classes: original, undersampled, oversampled dataset.
- Three "stress" classes: original, undersampled, oversampled, and both under/ over-sampled to arrive at the median class distribution.

During this analysis, we decided to use three target labels: 1 implying feelings of stress, 2 being neutral, and 3 representing a feeling of calmness. Prior to this decision, we trained several machine learning algorithms on the available data, but at the time, we lacked temperature inference which we believed to affect peoples' behavior and attitude towards daily activity. A stratified 10-fold CV is applied to the dataset and the accuracy computed is mean accuracy across them. Standard deviation is also computed across folds. The Kappa score is computed by comparing the original target class with the predicted target class after fitting the model on the training set (80:20 ratio with a random state of 42) and is used to evaluate the models. These results can be seen in Figs. 2 and 3.

		STRESS IS NOT IMPUTED				
		5 CLASSES		3 CLASSES		
		Original	Oversampled	Original	Oversampled	Undersampled
CART	Acc	27.29%,	50.57%,	37.42%,	41.93%,	36.66%,
	Std	0.0423,	0.0393,	0.0489,	0.0542,	0.0472,
	Kappa	0.2153	0.5583	0.1901	0.3437	-0.0443
RFC	Acc	37.94%,	69.86%,	43.25%,	52.31%,	42.19%,
	Std	0.0358,	0.03,	0.0431,	0.0337,	0.0563,
	Kappa	0.0271	0.7754	0.1584	0.5147	0.3071
GBC	Acc	36.49%,	54.42%,	44.47%,	46.4%,	42.72%,
	Std	0.0339,	0.0317,	0.0479,	0.0255,	0.0477,
	Kappa	0.1581	0.6004	0.2081	0.308	0.1495
XGB	Acc	35.38%,	67.54%,	44.68%,	51.2%,	43.99%,
	Std	0.0366,	0.0286,	0.0535,	0.0437,	0.0566,
	Kappa	0.2354	0.7948	0.2223	0.5104	0.2571

Fig. 2. Classifiers trained on StudentLife dataset without imputing the target class.

Considering a decision tree as a baseline model, we work with the StudentLife data where we have inferred temperature information from Meteostat. As aforementioned, we transform five levels of stress into three levels for trouble-free predictions, as well as not to over-manipulate the existing data since more than one class was a minority. We apply the following ensemble methods, all relying on decision trees:

- **Random Forest Classifier:** This method constructs a multitude of decision trees at training time and the predicted class is the mode of the classes of the individual trees. Decisions are made by each tree independently of the others, so the order of creation is irrelevant.
- **Gradient Boost Classifier:** Like random forests, gradient boosting is a set of decision trees, but instead of trees being built independently, it is an

		STRESS IS IMPUTED				
		5 CLASSES		3 CLASSES		
		Original	Oversampled	Original	Oversampled	Undersampled
CART	Acc	42.51%,	62.44%,	46.8%,	56.33%,	44.98%,
	Std	0.0161,	0.0192,	0.0256,	0.0449,	0.0477,
	Kappa	0.2378	0.7285	0.1966	0.4235	0.2308
RFC	Acc	53.47%,	81.62%,	57.99%,	70.62%,	55.34%,
	Std	0.0233,	0.0123,	0.0272,	0.0208,	0.0332,
	Kappa	0.2629	0.8999	0.3835	0.6349	0.395
GBC	Acc	56.85%,	62.02%,	59.13%,	59.31%,	54.26%,
	Std	0.0227,	0.0206,	0.0266,	0.0259,	0.0302,
	Kappa	0.4086	0.6975	0.4178	0.4627	0.4691
XGB	Acc	58.95%,	79.68%,	62.74%,	70.9%,	56.23%,
	Std	0.0277,	0.0162,	0.0226,	0.0143,	0.0336,
	Kappa	0.4177	0.8739	0.369	0.5748	0.4904

Fig. 3. Classifiers trained on StudentLife dataset with the target class imputed.

additive model which builds the trees in a forward stage-wise manner (sequentially).
- **XGBoost Classifier:** Unlike gradient boosting which utilizes the loss function of decision trees as a proxy for minimizing the error of the overall model, XGBoost uses the second-order derivative as an approximation. Additionally, it has advanced regularization which improves model generalization.

Models where the target variable was not imputed did not yield satisfactory results and are omitted from this paper. The rest of them are summarized in Fig. 4. Since we decided to work with an over-sampled dataset, Kappa Statistic is reported for those models. Moreover, the Kappa Statistic values are reported for the XGBoost Classifier as well because of its superior performance.

Confusion matrices of importance are those of the Random Forest Classifier and XGBoost Classifier. These are depicted in Fig. 5. It is clear that XGBoost shows greater performance, with more examples correctly classified.

Having encountered several difficulties in registering the XGB model for the application, we sought out an alternative. We make use of the Light Gradient Boosting Machine. This algorithm exploits a more novel technique of Gradient-based One-Side Sampling (GOSS) in order to filter data instances for computing the best split. Light GBM is a fast, distributed gradient boosting framework based on decision trees which requires lower memory than other boosting methods, making it an ideal fit for our task. It achieved accuracy of 65.12% (with a standard deviation of 0.0226 across folds), while maintaining a Kappa Statistic value of 0.578. The confusion matrix depicted in Fig. 6 an improvement in distinguishing between feelings of stress and calmness.

As a final step, we tell apart relevant features from irrelevant ones and order them with respect to the target variable by performing feature importance.

		STRESS IS IMPUTED			
		3 CLASSES			
		Original	Oversampled	Undersampled	Over/Under Sampled
CART	Acc	46.61%,	51.82%,	55.76%,	46.38%,
	Std	0.0471	0.0247,	0.03	0.0406
	Kappa		0.283		
RFC	Acc	54.61%,	65.28%,	62.43%,	53.44%,
	Std	0.0416	0.0255,	0.0285	0.0402
	Kappa		0.5704		
GBC	Acc	57.69%,	57.85%,	64.79%,	56.27%,
	Std	0.033	0.0214,	0.049	0.0309
	Kappa		0.403		
XGB	Acc	58.96%,	69.78%,	65.18%,	55.5%,
	Std	0.0381,	0.0229,	0.0328,	0.0469,
	Kappa	0.4361	0.6722	0.6393	0.497

Fig. 4. Classifiers trained on StudentLife dataset with the target class imputed.

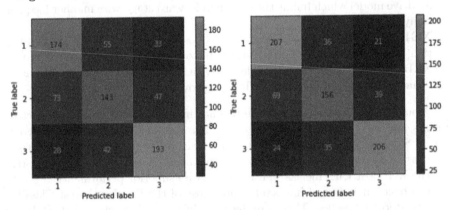

Fig. 5. Confusion matrices for the Random Forest and XGB classifiers.

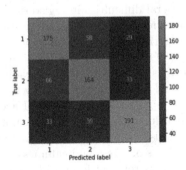

Fig. 6. Confusion matrix for the Light Gradient Boosting Machine algorithm.

After ranking the features, an error-curve for observing their contribution was plotted. Starting with the best ranked feature, our model's accuracy is plotted, and it proceeds to add features one by one, from highest to lowest, depending on the ranking. The error-curve clearly shows that all features we work with are important, therefore all of them are kept (Fig. 7).

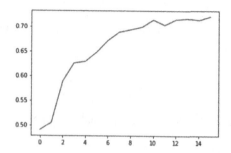

Fig. 7. Feature importance for the Light Gradient Boosting Machine algorithm.

5.2 Emotion Recognition Classifier

A sample architecture of a Convolutional Neural Network is given in Fig. 8, along with its unfolding version. With our implementation, we use four convolutional blocks. This model consists of four stages of convolutional and max pooling layers, followed by three fully connected layers comprising 256, 512 and 7 units respectively. We start with two convolutional layers, both with the same kernel size (3, 3), but with 32 and 64 units respectively. In the following stage, the convolutional layer uses 128 units with a kernel size of (5, 5). The last two convolution stages consist of 512 units with a filter of size (3, 3), each utilizing a L2 regularization for the kernel. After each layer of convolution applied, batch normalization is added to improve performance following a ReLU activation function prior to applying Max pooling with kernels of size (2, 2) and dropout with a rate set to 0.7. Batchnorm and dropout are too applied to the fully connected layer blocks, except for the last one where a softmax activation is used. The initial learning rate, weight decay rate and batch size are set to 0.0001, $1e^{-6}$ and 256, respectively. We train our model for 60 epochs. The learning rate is factored by 0.2 if the validation loss does not improve for 6 epochs. To additionally protect the network from overfitting, early stopping is applied by monitoring the validation loss for training to be stopped after 3 epochs with no improvement.

With our combination of features, the results obtained are satisfactory and provide further incentive to show that less complex models can go toe-to-toe with larger architectures [28,29]. Evaluating the model, we can conclude that the parameters we set during training performed well enough, with accuracy

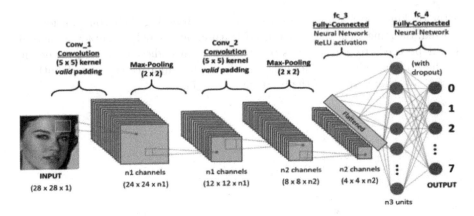

Fig. 8. A sample Convolutional Neural Network.

of 76.35% obtained during training and 67.44% during validation as shown in Fig. 9. For an error analysis, readers may address the confusion matrix in Fig. 10, from which it is clear that the model fails to recognize the feeling of disgust. We can not say that it was unexpected, since indeed disgust is the minority class in this dataset, with a support of only 436 and 111 images in the training and validation set, respectively.

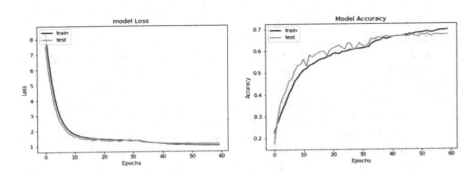

Fig. 9. CNN model loss and accuracy.

Remark 1. If needed.

6   Application Development

While developing the application the focus was on covering all the essential functionalities required for seamless monitoring and production of data. User activity, accelerometer, location, phone lock, illumination and phone charge are

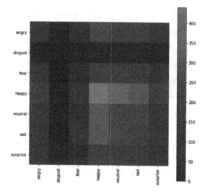

Fig. 10. Confusion matrix for the Emotion Recognition model.

all monitored by a common set of sensors and data sources on both Android and iPhone. The accelerometer is most often used to automatically change the screen orientation depending on the device physical position, but it can also track user activity. Physical activity and depression levels are shown to be significantly linked in a previous study [43]. The necessary sensors are supported by both iPhone and Android devices. In reference to other papers thirty-one studies used the Android operating system (OS), compared to two that used Apple iOS [44,45]. This could be explained by the access granted on Android phones, making it easier for data capture, communication, and processing tasks to run in the background. In contrast, Apple's iOS made it harder for applications to access data from other applications without explicit user permission [46]. However, our application is compatible with both operating systems.

Upon installment, "Emphasize" requires permission for the services and initiates the application to begin continuous sampling of the sensor data. iPhone devices are data sampling once every 3 s for light, lock and charge and every 30 s for GPS coordinates. Regarding the collection of the activity data on iPhone, the application is not sampling data every once a while, but it uses an instance method for querying activity data for the specified time period i.e., the previous 24 h. For inference of physical activities and location services on Android, the application relies on Google play services, namely, two APIs were used: Activity Recognition API and Fused Location Provider API. Android devices receive updates on activity states as soon as they are available, whereas for location data three intervals have been defined: update interval, which is set to sixty seconds, fastest update interval, set to thirty seconds and maximum waiting time of three minutes. The sampling rates were set in order to maximize the battery life, and it makes no difference because the focus is on the overall behavior, not on analyzing short, fine grained movements. This process, which involves the continuous sampling of the sensor data, lasts until either the battery is drained or the user terminates the application.

Each light cycle has a start and an end timestamp which is generated if the time difference i.e. the period spent in a dark environment exceeds one minute, as opposed to the sensors in the StudentLife data set which report duration longer than one hour. Monitoring ambient light and reporting time spent in a dark environment is done with the light sensor. This is implemented with a foreground service which starts running when the user logs in to the application and continues to do so even when the application is sent to the background. The foreground service implements a sensor event listener that captures the changes of light intensity. Two more foreground services are needed, one which refers to the locking state of the device and another which refers to its charging state. The same method with the timestamps measurements is used for lock duration and charge duration. It is worth mentioning that the states of locking and charging present system events and thus, these actions are broadcast system-wide. Each broadcast receiver captures an action that describes the change of state, such as phone lock, unlock, put to charging or taken off charger. Furthermore, there is no calculation of duration between states and all occurrences are taken into account. A high-level description of its sensing system is shown in Fig. 11.

Data is sent by answering a survey which contains a few questions related to activity, sleep and stress. Furthermore, there is an option of taking a picture of one's face and uploading it together with the survey. For this task, the emotion recognition model described in Sect. 5 is used. In addition to stress predictions, the users who will make use of this functionality can also receive reports on their mood throughout the month.

The user interface consists of six views. Aside from the standard login and registration user interface, "Emphasize" consists of a home view, survey view and tracking of predictions. The home view is made up of basic information about the logged user whereas the survey view features the survey to be answered as well as the option for sending a photo, as seen in Fig. 12.

The view which serves as a diary of past predictions incorporates a calendar view and a pie chart associated with the user's mood generated by the emotion classifier. Each day in the calendar is colored with respect to the predicted stress level. This is shown in Fig. 13.

6.1 Sensing Data

The data is collected without any user interaction whereas the process of sending the data is left to the user, as the goal was for the application to act as a daemon which collects the daily data of the user without the need of user input. After the data is collected using the aforementioned APIs it is stored in arrays on the phone until the questionnaire is completed, at which point it is sent to the server in a form of a JSON object through a REST request where it is stored securely on the web server.

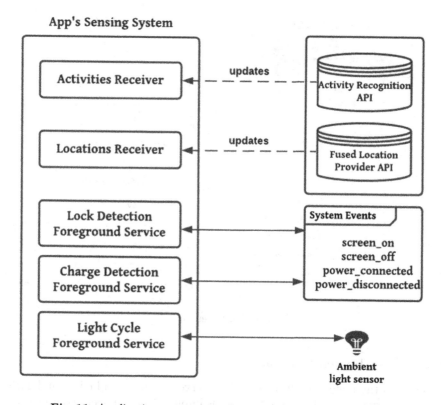

Fig. 11. Application components for gathering the sensing data.

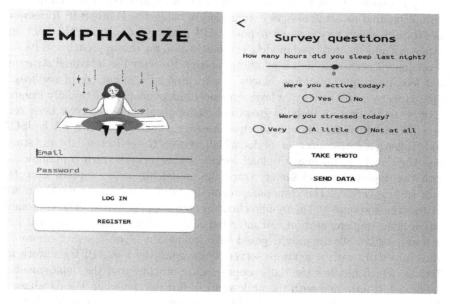

Fig. 12. Login. Survey.

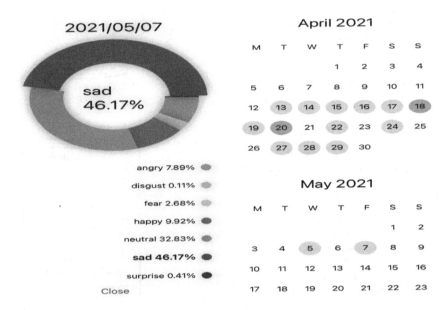

Fig. 13. Calendar.

6.2 Server Implementation

Aside from the mobile applications which serve as a front-end client and gather the necessary sensory data, a web server is implemented for secure data storage, data prepossessing and model deployment. From an architectural standpoint it is implemented as a microservice architecture using the FeathersJS framework for NodeJS which serves as an exposed API gateway connecting the back-end data store and the deployed model. The motivation for this approach was its fast development cycle and high level of modularity. Presented as a layered structure and loosely coupled building blocks, constituted of a summary of services it allowed us to craft the service layer, protocol independent, functionally isolated in order to facilitate a scalable growth of our application. The server layer consists of a authentication service which uses local authentication with JWT (JSON Web Token), a data collection service which receives the collected data and stores it in the database, an image upload service and a service which communicates with the deployed model server. Once the model has made a prediction, the data service receives the response, stores the predictions in the database and returns the response to the mobile client in an adequate format. The communication layer implementation can interchangeably act as a REST endpoint and as a web socket. As the only exposed entry point, the NodeJS server communicates internally with a separate server, built using the FastAPI framework for Python, which handles the data preprocessing pipeline and the deployment of the model. Continuing with the philosophy of fast development, the database of choice was MongoDB due to its flexible data storage structure, which only communicates with the NodeJS server and is used for the storage of the sensor data, the predictions made and as a bucket storage for the images used for the emotion recognition. A simple overview is shown in Fig. 14. To have a more granular control over the life-cycle of the model, the MLflow platform was implemented. Using its Tracking Server API which aside from its function as a singular source of model data also tracks all the runs the model had during its training phase, allowing us to record and track the progress as well as version our model further on. All of the model data necessary to track and deploy the model is stored in an SQLite database. The server is exposed to the outside world using NGINX which acts as a http reverse proxy server in charge of sanitizing the incoming http requests and rerouting them to our server. With regards to the preprocessing pipeline, the data is cleaned, aggregated and transformed into a pandas data frame before its fed to the model, which aside from the sensory data consist of weather data which is accessed using the Meteostat Python library.

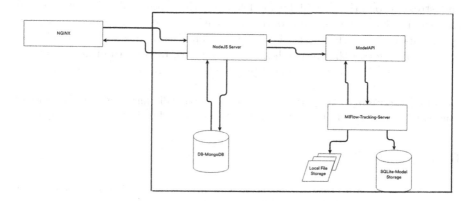

Fig. 14. Server diagram.

7   Concluding Remarks

The critical factor associated with stress is in its chronicity. Chronic stressors include daily hassles, frustration of traffic jams, work overload, financial difficulties, marital arguments or family problems. As we build up anger inside ourselves toward any of these situations, or the guilt and resentment we hold toward others and ourselves, all result in the production of the same effects on the hypothalamus [47]. Instead of discharging this stress, however, we hold it inside where its effects become cumulative. Living in times when stress and anxiety are reaching an all-time high, especially in younger generations, we strive to provide ways in combating stress and promoting well being.

In this paper, it is demonstrated how passive mobile sensing can effectively portray stress by collecting data in an unstructured environment with an unknown source or number of stressors. The architecture of the "Emphasize" application is presented and it is shown how mobile phones can act as stress detectors.

The results achieved are slightly better than the results found during literature review, the difference in our work being that we exclude audio data for privacy reasons whereas we build an emotion recognition classifier. General models have lower performance, but they do not require user-specific labeled data which is time-costly gather and analyze. User-specific models would perform best since they are personalized for each individual user, but they require more labeled data and thus, longer times for collection. A possibility is to construct a model based on similarity among users, where users would be grouped together based on their behaviour. Conversely, one of the major drawbacks we faced was the lack of available mobile sensing data which resulted in the use of imputations and oversampling methods. However, once the application is deployed, an adequate amount of data with superior quality will be able to be attained.

Further research would concentrate on implementing reinforcement learning to allow for improved future predictions. Moreover, how emotion recognition and stress recognition could be modeled together would also be investigated thoroughly. Since this is a prototype application, there still exist some serious optimization issues that must be resolved. This paper presents a number of opportunities for further research in multiple areas, ranging from mental health to mobile sensing and the mobile applications domain.

Acknowledgement. The work in this paper was partially financed by the Faculty of Computer Science and Engineering, Ss. Cyril and Methodius University in Skopje.

References

1. Mehrotra, A., Musolesi, M.: Designing effective movement digital biomarkers for unobtrusive emotional state mobile monitoring (2017)
2. Harari, G., et al.: An evaluation of students' interest in and compliance with self-tracking methods: recommendations for incentives based on three smartphone sensing studies. Soc. Psychol. Pers. Sci. **8**(5), 479–492 (2017). https://doi.org/10.1177/1948550617712033
3. Harari, G., Wang, W., Müller, S., Wang, R., Campbell, A.: Participants' compliance and experiences with self-tracking using a smartphone sensing app (2017)
4. Torous, J., Staples, P., Onnela, J.-P.: Realizing the potential of mobile mental health: new methods for new data in psychiatry. Curr. Psychiatry Rep. **17**(8), 1–7 (2015). https://doi.org/10.1007/s11920-015-0602-0
5. Levine, L.M., et al.: Anxiety detection leveraging mobile passive sensing. In: Alam, M.M., Hämäläinen, M., Mucchi, L., Niazi, I.K., Le Moullec, Y. (eds.) BODYNETS 2020. LNICST, vol. 330, pp. 212–225. Springer, Cham (2020). https://doi.org/10.1007/978-3-030-64991-3_15
6. Wang, R., et al.: StudentLife: assessing mental health, academic performance and behavioral trends of college students using smartphones. In: UbiComp 2014 - Proceedings of the 2014 ACM International Joint Conference on Pervasive and Ubiquitous Computing (2014)
7. Ben-zeev, D., Scherer, E., Wang, R., Xie, H., Campbell, A.: Next-generation psychiatric assessment: using smartphone sensors to monitor behavior and mental health. Psychiatr. Rehabil. J. **38**, 218 (2015)

8. Haim, S., Wang, R., Lord, S., Loeb, L., Zhou, X., Campbell, A.: The mobile photographic stress meter (MPSM) (2015)
9. Wang, R., et al.: CrossCheck: toward passive sensing and detection of mental health changes in people with schizophrenia (2016)
10. Henningsen, G.M., et al.: Measurement of salivary immunoglobulin A as an immunologic biomarker of job stress. Scand. J. Work Environ. Health **18**, 133–136 (1992)
11. Horvitz, E., Apacible, J.: Learning and reasoning about interruption. In: ICMI 2003: Fifth International Conference on Multimodal Interfaces, pp. 20–27 (2003)
12. Horvitz, E., Jacobs, A., Hovel, D.: Attention-sensitive alerting (2013)
13. Horvitz, E., Kadie, C., Paek, T., Hovel, D.: Models of attention in computing and communication: from principles to applications. Commun. ACM **46**, 52–59 (2003)
14. Horvitz, E., Koch, P., Apacible, J.: BusyBody: creating and fielding personalized models of the cost of interruption. In: Proceedings of the ACM Conference on Computer Supported Cooperative Work, CSCW, pp. 507–510 (2004)
15. Guidoux, R., et al.: A smartphone-driven methodology for estimating physical activities and energy expenditure in free living conditions. J. Biomed. Inform. **52**, 271–278 (2014)
16. Al-mardini, M., Aloul, F., Sagahyroon, A., Al-husseini, L.: Classifying obstructive sleep apnea using smartphones. J. Biomed. Inform. **52**, 251–259 (2014)
17. Osmani, V.: Smartphones in mental health: detecting depressive and manic episodes. IEEE Pervasive Comput. **14**, 10–13 (2015)
18. Saeb, S., et al.: Mobile phone sensor correlates of depressive symptom severity in daily-life behavior: an exploratory study. J. Med. Internet Res. **17**, e4273 (2015)
19. Harari, G., Gosling, S., Wang, R., Chen, F., Chen, Z., Campbell, A.: Patterns of behavior change in students over an academic term: a preliminary study of activity and sociability behaviors using smartphone sensing methods. Comput. Hum. Behav. **67**, 129–138 (2017)
20. Bakker, J., Pechenizkiy, M., Sidorova, N.: What's your current stress level? Detection of stress patterns from GSR sensor data. In: Proceeding of the 2nd Hacdais Workshop Collocated with IEEE, ICDM 2011, pp. 573–580 (2011)
21. Poh, M., Swenson, N., Picard, R.: A wearable sensor for unobtrusive, long-term assessment of electrodermal activity. IEEE Trans. Biomed. Eng. **57**, 1243–52 (2010)
22. Lu, H., et al.: StressSense: detecting stress in unconstrained acoustic environments using smartphones. In: UbiComp 2012 - Proceedings of the 2012 ACM Conference on Ubiquitous Computing, pp. 351–360 (2012)
23. Sano, A., Picard, R.: Stress recognition using wearable sensors and mobile phones. In: Proceedings - 2013 Humaine Association Conference on Affective Computing and Intelligent Interaction, ACII 2013, pp. 671–676 (2013)
24. Cohen, S.: Perceived Stress Scale. Mind Garden Inc. (1994)
25. Chawla, N., Bowyer, K., Hall, L., Kegelmeyer, W.: SMOTE: synthetic minority over-sampling technique. J. Artif. Intell. Res. (JAIR) **16**, 321–357 (2002)
26. Mitchell, T.: Machine Learning. McGraw-Hill (1997)
27. Goodfellow, I., et al.: Challenges in representation learning: a report on three machine learning contests. Neural Netw. **64**, 59–63 (2013)
28. Georgescu, M., Ionescu, R., Popescu, M.: Local learning with deep and handcrafted features for facial expression recognition. IEEE Access **7**, 64827–64836 (2019)
29. Fathallah, A., Abdi, L., Douik, A.: Facial expression recognition via deep learning (2017)
30. Shiffman, S., Stone, A., Hufford, M.: Ecological momentary assessment. Annu. Rev. Clin. Psychol. **4**, 1–32 (2008)

31. Ratner, B.: The correlation coefficient: its values range between +1/-1, or do they? J. Target. Measur. Anal. Mark. **17**, 139–142 (2009). https://doi.org/10.1057/jt.2009.5

32. Morin, C., Rodrigue, S., Ivers, H.: Role of stress, arousal, and coping skills in primary insomnia. Psychosom. Med. **65**, 259–67 (2003)

33. Pillai, V., Roth, T., Mullins, H., Drake, C.: Moderators and mediators of the relationship between stress and insomnia: stressor chronicity, cognitive intrusion, and coping. Sleep **37**(7), 1199–1208A (2014)

34. Sadeh, A., Keinan, G., Daon, K.: Effects of stress on sleep: the moderating role of coping style. Health Psychol. Off. J. Div. Health Psychol. Am. Psychol. Assoc. **23**, 542–5 (2004)

35. Zoccola, P., Dickerson, S., Lam, S.: Rumination predicts longer sleep onset latency after an acute psychosocial stressor. Psychosom. Med. **71**, 771–775 (2009)

36. Bonnet, M.H., Arand, D.L.: Situational insomnia: consistency, predictors, and outcomes. Sleep **26**, 1029–1036 (2004)

37. Saeb, S., Cybulski, T., Kording, K., Mohr, D.: Scalable passive sleep monitoring using mobile phones: opportunities and obstacles. J. Med. Internet Res. **19**, e118 (2017)

38. Ciman, M., Wac, K.: iSenseSleep: smartphone as a sleep duration sensor (preprint). JMIR Mhealth Uhealth **7**, 7 (2018)

39. Henry, D., Tolan, P., Gorman-Smith, D.: Cluster analysis in family psychology research. J. Fam. Psychol. JFP: J. Div. Fam. Psychol. Am. Psychol. Assoc. (Division 43) **19**, 121–32 (2005)

40. Rousseeuw, P.J.: Silhouettes: a graphical aid to the interpretation and validation of cluster analysis. J. Comput. Appl. Math. **20**, 53–65 (1987)

41. Pedregosa, F., et al.: Scikit-learn: machine learning in Python. JMLR **12**, 2825–2830 (2011)

42. The Weather's Record Keeper (n.d.). https://www.meteostat.net/en/. Accessed 15 Apr 2021

43. Fox, K.: The influence of physical activity on mental well-being. Public Health Nutr. **2**, 411–418 (1999)

44. Bhat, S., et al.: Is there a clinical role for smartphone sleep apps? Comparison of sleep cycle detection by a smartphone application to polysomnography. J. Clin. Sleep Med. JCSM **11**, 709–715 (2015). Official Publication of the American Academy of Sleep Medicine

45. Natale, V., Drejak, M., Erbacci, A., Tonetti, L., Fabbri, M., Martoni, M.: Monitoring sleep with a smartphone accelerometer. Sleep Biol. Rhythms **10**(4), 287–292 (2012). https://doi.org/10.1111/j.1479-8425.2012.00575.x

46. Cornet, V., Holden, R.: Systematic review of smartphone-based passive sensing for health and wellbeing. J. Biomed. Inform. **77**, 120–132 (2017)

47. Salleh, M.: Life event, stress and illness. Malays. J. Med. Sci. MJM **15**, 9–18 (2008)

Author Index

Printed in the United States
by Baker & Taylor Publisher Services